Flow and Heat Transfer Research in Multiphase Flow and Porous Media

Flow and Heat Transfer Research in Multiphase Flow and Porous Media

Editor

Liangxing Li

MDPI

Basel • Beijing • Wuhan • Barcelona • Belgrade • Novi Sad • Cluj • Manchester

Editor
Liangxing Li
Xi'an Jiaotong University
Xi'an
China

Editorial Office
MDPI
St. Alban-Anlage 66
4052 Basel, Switzerland

This is a reprint of articles from the Special Issue published online in the open access journal *Applied Sciences* (ISSN 2076-3417) (available at: https://www.mdpi.com/journal/applsci/special_issues/heattransfer_multiphaseflow_porousmedia).

For citation purposes, cite each article independently as indicated on the article page online and as indicated below:

Lastname, A.A.; Lastname, B.B. Article Title. *Journal Name* **Year**, *Volume Number*, Page Range.

ISBN 978-3-7258-1139-7 (Hbk)
ISBN 978-3-7258-1140-3 (PDF)
doi.org/10.3390/books978-3-7258-1140-3

Contents

About the Editor

Liangxing Li

Dr. Liangxing Li is an Associate Professor at the State Key Laboratory of Multiphase Flow in Power Engineering (MPFL), Xi'an Jiaotong University (XJTU), China. He received his Ph.D. in Power Engineering and Engineering Thermophysics in 2008 from XJTU. Then, he worked as a Postdoc. and Senior Researcher from 2008 to 2012 at the Physical Department of the Royal Institute of Technology, Sweden. Hereafter, he worked as an Associate Professor at MPFL from 2013. His research is involved in the field of multiphase flow and heat transfer, transport in porous media, flow boiling and heat transfer, nuclear power safety, and thermal hydraulics in nuclear reactors. He has published more than 100 peer-reviewed articles and two books, and he has been granted more than 10 patents and two software copyrights.

applied sciences

MDPI

Editorial

Latest Progress and Applications of Multiphase Flow and Heat Transfer

Liangxing Li

State Key Laboratory of Multiphase Flow in Power Engineering, Xi'an Jiaotong University, Xi'an 710049, China; liangxing.li@xjtu.edu.cn

1. Introduction

Multiphase flow and heat transfer are critical in both traditional and emerging area of engineering research [1] and are widely involved in industrial processes and academic studies [2,3], including petrochemical engineering [4,5], chemical engineering [6,7], nuclear engineering [8–10], mechanical engineering [11,12], and ocean engineering [13,14], as well as renewable energy [15,16], energy storage [17,18], and cleaning coal technology [19,20], etc. With the rapid development of various interdisciplinary subjects and technologies, the novel technologies and innovative applications of multiphase flow and heat transfer are proposed and developed in an endless stream and show great potential. On the other hand, studies of multiphase flow and heat transfer commonly couple various physical processes, such as phase change [21,22], interface evolution and interaction, physical or chemical reactions, multiscale coupling analysis, spatiotemporal transient dynamics, and multicomponent flow and mass transfer. Although a lot of studies and great efforts have been performed to understand the complicated multiphase flow and heat transfer phenomena and to reveal new mechanisms and theories, there are still many issues that need to be clarified from both theoretical and applied aspects of this important field [1]. Correspondingly, the methods, algorithms, and modeling of numerical simulation also pose great challenges with regard to accurately capturing and predicting the heat transfer behavior of multiphase flow.

This Special Issue aims to provide a top-notch platform to introduce the latest progress and various applications in the area of multiphase flow and heat transfer. The scope of this Special Issue includes the aspects of theoretical derivation and analysis, model development and simulation, and experimental investigation and engineering applications, with hopes to be invaluable for scientists and researchers interested in multiphase flows. Generally, seven articles in this Special Issue cover the latest research advances in academia (mainly from the universities or institutes) and industry, such as KBR in the US and CNNC in China. The contributed scientists are from nine countries, including the USA, Russia, Germany, Belgium, Serbia, Australia, Colombia, Japan, and China. Moreover, I would like to express my gratitude to all authors who have contributed to this Special Issue.

2. An Overview of Published Articles

The first article by Mr. Li et al. from the Northeastern University of China and Prof. Tu from the Royal Melbourne Institute of Technology (RMIT) University of Australia conducted a numerical study on spontaneous steam condensation in a nozzle. A wet steam model with entropy generation rates was proposed in the study, considering three different mechanisms, including viscous dissipation, heat transport, and phase change. Different geometric parameters of the nozzle (like throat radius, divergent section expansion angle, and divergent section length) were designed, and the effects on the spontaneous steam condensation were discussed. In addition, the performance of the nozzle was also studied by comparing the mass flow rate, total entropy production, and liquid mass fraction. The calculated results demonstrate that increasing the throat diameter or reducing the expansion

Citation: Li, L. Latest Progress and Applications of Multiphase Flow and Heat Transfer. *Appl. Sci.* **2024**, *14*, 3369. https://doi.org/10.3390/app14083369

Received: 11 April 2024
Accepted: 16 April 2024
Published: 17 April 2024

angle of the divergent section restrains steam condensation. The irreversible energy caused by steam condensation results in energy waste and weakens the nozzle's performance.

The second article is also an international collaborative work related to the two-phase flow in petroleum engineering. Researchers from Asia (Saudi Arabia and Kuwait), Europe (Russia, Serbia, Germany, Belgium), and South America (Colombia) propose a non-invasive method to identify the flow pattern and to determine the volume percentage of two-phase flow in a scale-laden petroleum pipeline by employing artificial intelligence technologies. In the tested petroleum pipeline, a dual-energy gamma source emits photons, and two detectors are placed in front of the gamma source at a 45-degree angle to the center of the pipe: one is for transmitted photons and the other one is for scattered photons. The Monte Carlo N-Particle code (MCNP) is employed to analyze the flow patterns at different volume percentages as well as to calculate the scale thickness inside the pipe. The feature extraction technique of the time domain is adapted to extract time characteristics, like skewness, kurtosis, and 4th-order moment. Finally, the flow patterns and the volume percentages are identified by the two designed multilayer perceptron neural networks.

Researchers of the third article are from Japan (Tokyo Institute of Technology and Tokyo University of Agriculture and Technology) and Indonesia (Institut Teknologi Nasional Bandung), focusing on the hydrodynamic fingering induced by gel formation. They conduct Hele–Shaw cell displacement experiments for a miscible fluid system by using skim milk and an aqueous citric acid solution. The effects of gel film formation on fingering instability are discussed and investigated, and a mathematical model is developed for the sequential growth of gel film formation at the fingertip. The results show that mixing skim milk with the aqueous citric acid solution causes the formation of gel film, which leads to interface instability. The diffusion of citric acid causes the formation of gel film to thicken over time. As the flow increases, the width of the fingers remains constant while the finger number increases linearly before the fingers merge. Based on the interaction between the diffusion of citric acid and the elongation of the fingertip, a mathematical model of sequential film thickness growth is developed for a bubble-like fingertip structure. The model is helpful for providing us with a better understanding of the fundamental growth mechanism of the bubble-like fingertip.

The fourth article written by scholars from academia (Louisiana State University and University of Texas at Austin) and industry (KBR) in the USA focuses on the influence of porous media on the interaction between wave-induced sloshing and the dynamics of the floating body, which is very important to the structural vibration control in civil and ocean engineering. A numerical algorithm that couples fluid computation (for sloshing fluid and ambient waves) and rigid-body dynamics (for the floating platform) is proposed. The Eulerian–Lagrangian method is employed to calculate the hydrodynamic in both the pore-flow domain and the pure-water domain. The Newmark time integration method is adopted for the rigid body dynamics to ensure numerical stability. The modeling freedom of the sloshing fluid is reduced, and the numerical process is fast and inexpensive. According to the frequency response analysis, the effectiveness of the porous media is verified for reducing the vibration and mitigating the sloshing response. It is believed that the porous media will reduce the hydrodynamic pressure and improve the integrity of the liquid container.

In the fifth article, Li et al. from the Xi'an Jiaotong University of China develop a high-accuracy flow pattern identification method for air–water two-phase flow in the porous media by combining signal feature extraction technologies and machine learning technologies. According to the differential pressure signals of two-phase flow in porous beds, three parameters related to time domain characteristics, including the mean, standard deviation, and range of the signals, are analyzed and extracted using a statistical method. In addition, the time–frequency domain features of the signals are also extracted with the empirical mode decomposition (EMD) method. Then, machine learning technologies, such as the support vector machine (SVM) and BP neural network, are adopted to train and construct different flow pattern identification models based on extracted signal feature parameters. In the end, an online and high-accuracy intelligent system is established to identify the flow patterns in porous media.

The sixth article focuses on the enhanced heat transfer characteristics of spherical heat storage related to phase change thermal storage technology. Researchers from the Inner Mongolia Agricultural University and Inner Mongolia Agricultural University of China designed and constructed a test system of melting spherical heat storage units to experimentally study the melting characteristics of a CuO-paraffin wax composite in a spherical heat storage unit. The influences of parameters, including the pin fin numbers, the temperature in the water bath, and the CuO nanoparticles in paraffin, are discussed and analyzed. In addition, a regression model is fitted to investigate the effects of different parameters on the melting time of the phase change material (PCM). The results show that the interaction between the water bath temperature and the pin fin numbers significantly affects the melting time of the PCM in the heat storage unit. A prediction model is further established to calculate the melting time via the response surface methodology. The work can prove to be helpful in the design of thermal storage units.

The seventh article focuses on the process of steam condensation heat transfer in nuclear engineering. Li et al. from China Nuclear Power Engineering Co. Ltd. propose a comprehensive heat transfer model (considering liquid film heat transfer, steam condensation, and convective heat transfer), aiming to study the heat transfer process of steam condensation with non-condensable gas. A model is then adopted via the Integrated Program of Severe Accident Analysis (PISAA) for nuclear power plants. By comparing the calculated results from the PISAA program with those obtained using the traditional containment analysis codes of nuclear reactors, as well as the experimental data in the Wisconsin condensation tests, the model is verified and shows good performance with a stable calculation process, less iteration, and fast convergence. Moreover, sensitivity analysis of the heat transfer coefficient parameters shows that the average error is about 10% for the condensation heat transfer coefficient and that the maximum value is below 30%, which indicates that the model is suitable for thermal-hydraulic analysis in nuclear engineering.

3. Conclusions

The compilation of studies in this Special Issue is devoted to highlighting the recent progress and various applications of multiphase flow and heat transfer, ranging from conventional research and interdisciplinary research; additionally, it aims to introduce the different methodologies and interdisciplinary manners that are employed in the respective case studies.

In summary, the first paper and the seventh paper both focus on the phenomenon of steam condensation, while the numerical simulation performed in the first paper and the seventh paper establishes a calculation model to analyze the steam condensation heat transfer with non-condensable gas; then, the model is employed using the analysis code for severe accidents of nuclear reactors. Applications of machine learning technologies in multiphase flow and heat transfer are discussed and explored in the second and the fifth articles, corresponding to petroleum engineering and nuclear engineering, respectively. Based on the experimental studies conducted in the third paper, a mathematical model is established to better understand the fundamental growth mechanism of the bubble-like fingertip. The fourth paper proposes a numerical algorithm used in ocean engineering, while the sixth paper conducts an experimental study related to the field of energy storage.

In the end, I would like to emphasize that the studies compiled in this Special Issue may be seen as a starting point to inspire scientists and researchers in the field of multiphase flow and heat transfer to explore innovative studies, including the fundamentals of multiphase flow and heat transfer, interdisciplinary studies, those under extreme working conditions, etc.

Conflicts of Interest: The authors declare no conflict of interest.

List of Contributions

1. Li, H.; Wang, X.; Huang, H.; Ning, J.; Tu, J. A Numerical Analysis of the Influence of Nozzle Geometric Structure on Spontaneous Steam Condensation and Irreversibility in the Steam Ejector Nozzle, *Appl. Sci.* **2021**, *11*, 11954. https://doi.org/10.3390/app112411954.
2. Alanazi, A.K.; Alizadeh, S.M.; Nurgalieva, K.S.; Nesic, S.; Grimaldo Guerrero, J.W.; Abo-Dief, H.M.; Eftekhari-Zadeh, E.; Nazemi, E.; Narozhnyy, I.M. Application of Neural Network and Time-Domain Feature Extraction Techniques for Determining Volumetric Percentages and the Type of Two Phase Flow Regimes Independent of Scale Layer Thickness, *Appl. Sci.* **2022**, *12*, 1336. https://doi.org/10.3390/app12031336.
3. Nasir, M.; Yamaguchi, R.; She, Y.; Patmonoaji, A.; Mahardika, M.A.; Wang, W.; Li, Z.; Matsushita, S.; Suekane, T. Hydrodynamic Fingering Induced by Gel Film Formation in Miscible Fluid Systems: An Experimental and Mathematical Study. *Appl. Sci.* **2022**, *12*, 5043. https://doi.org/10.3390/app12105043.
4. Tsao, W.H.; Chen, Y.C.; Kees, C.E.; Manuel, L. The Effect of Porous Media on Wave-Induced Sloshing in a Floating Tank. *Appl. Sci.* **2022**, *12*, 5587. https://doi.org/10.3390/app12115587.
5. Li, X.; Li, L.; Wang, W.; Zhao, H.; Zhao, J. Machine Learning Techniques Applied to Identify the Two-Phase Flow Pattern in Porous Media Based on Signal Analysis. *Appl. Sci.* **2022**, *12*, 8575. https://doi.org/10.3390/app12178575.
6. Lu, L.; Tian, R.; Gong, X.; Zhao, Y. Enhanced Heat Transfer Study of Spherical Heat Storage Based on Response Surface Methodology. *Appl. Sci.* **2023**, *13*, 8595. https://doi.org/10.3390/app13158595.
7. Li, H.; Yang, X.; Wang, C.; Shi, S.; Ma, R.; Yuan, Y. Research and Application of Steam Condensation Heat Transfer Model Containing Noncondensable Gas on a Wall Surface. *Appl. Sci.* **2023**, *13*, 10520. https://doi.org/10.3390/app131810520.

References

1. Chen, L.; Ghajar, A. Frontiers and progress in multiphase flow and heat transfer. *Heat Transf. Eng.* **2019**, *40*, 1299–1300. [CrossRef]
2. Faghri, A.; Zhang, Y. *Fundamentals of Multiphase Heat Transfer and Flow*; Springer Nature: Cham, Switzerland, 2020; ISBN 978-3-030-22137-9. [CrossRef]
3. Flint, T.F.; Smith, M.C.; Shanthraj, P. Magneto-hydrodynamics of multi-phase flows in heterogeneous systems with large property gradients. *Sci. Rep.* **2021**, *11*, 18998. [CrossRef] [PubMed]
4. Arastoopour, H.; Gidaspow, D.; Lyczkowski, R.W. *Transport Phenomena in Multiphase Systems*; Springer Nature: Cham, Switzerland, 2022; ISBN 978-3-030-68578-2. [CrossRef]
5. Zhang, J.; Yuan, H.; Zhao, J.; Mei, N. Theoretical and experimental investigations of identifying the ingredients of an oil-water mixture based on a characteristic fluid inverse problem. *Int. J. Thermophys.* **2016**, *37*, 128. [CrossRef]
6. Bracconi, M. CFD modeling of multiphase flows with detailed microkinetic description of the surface reactivity. *Chem. Eng. Res. Des.* **2022**, *179*, 564–579. [CrossRef]
7. Kumar, V.; Nigam, V.K.D.P. Multiphase fluid flow and heat transfer characteristics in microchannels. *Chem. Eng. Sci.* **2017**, *169*, 34–66. [CrossRef]
8. Jiang, S.; Tu, J.; Yang, X.; Gui, N. *Multiphase Flow and Heat Transfer in Pebble Bed Reactor Core*; Springer: Singapore, 2020; ISBN 978-981-15-9565-3. [CrossRef]
9. Zhang, Z.; Li, L.; Ma, W.; Yuan, Y.; Yang, X.; Ma, R. Experimental and numerical studies on the two-dimensional flow characteristics in the radially stratified porous bed. *Int. Commun. Heat Mass Transf.* **2022**, *133*, 105940. [CrossRef]
10. Li, L.; Zou, X.; Wang, H.; Zhang, S.; Wang, K. Investigations on two-phase flow resistances and its model modifications in a packed bed. *Int. J. Multiph. Flow* **2018**, *101*, 24–34. [CrossRef]
11. Pandey, A.; Madduri, B.; Perng, C.; Srinivasan, C.; Dhar, S. Multiphase flow and heat transfer in an electric motor. In Proceedings of the ASME 2022 International Mechanical Engineering Congress and Exposition, Columbus, OH, USA, 30 October–3 November 2022; Volume 8: Fluids Engineering; Heat Transfer and Thermal Engineering; American Society of Mechanical Engineers: New York, NY, USA, 2022. [CrossRef]
12. Tong, F.; Jing, L.; Zimmerman, R.W. A fully coupled thermo-hydro-mechanical model for simulating multiphase flow, deformation and heat transfer in buffer material and rock masses. *Int. J. Rock Mech. Min. Sci.* **2010**, *47*, 205–217. [CrossRef]
13. Song, W.; Li, W.; Lin, S.; Zhou, X.; Han, F. Flow pattern evolution and flow-induced vibration response in multiphase flow within an M-shaped subsea jumper. *Ocean Eng.* **2024**, *298*, 117213. [CrossRef]
14. Seo, Y.S.; Chung, S.M.; Park, J.C. Multiphase-thermal flow simulation in a straight vacuum-insulated LH2 pipe: Cargo handling system in LH_2 carrier. *Ocean Eng.* **2024**, *297*, 117030. [CrossRef]
15. Feng, G.; Wang, Y.; Xu, T.; Wang, F.; Shi, Y. Multiphase flow modeling and energy extraction performance for supercritical geothermal systems. *Renew. Energy* **2021**, *173*, 442–454. [CrossRef]

16. Khan, R.; Ullah, S.; Qahtani, F.; Pao, W.; Talha, T. Experimental and numerical investigation of hydro-abrasive erosion in the Pelton turbine buckets for multiphase flow. *Renew. Energy* **2024**, *222*, 119829. [CrossRef]
17. Rostami, S.; Afrand, M.; Shahsavar, A.; Sheikholeslami, M.; Kalbasi, R.; Aghakhani, S.; Shadloo, M.S.; Oztop, H.F. A review of melting and freezing processes of PCM/nano-PCM and their application in energy storage. *Energy* **2020**, *211*, 118698. [CrossRef]
18. Shi, X.J.; Zhang, P. conjugated heat and mass transfer during flow melting of a phase change material slurry in pipes. *Energy* **2016**, *99*, 58–68. [CrossRef]
19. Ou, Z.; Jin, H.; Ren, Z.; Zhu, S.; Song, M.; Guo, L. Mathematical model for coal conversion in supercritical water: Reacting multiphase flow with conjugate heat transfer. *Int. J. Hydrogen Energy* **2019**, *44*, 15746–15757. [CrossRef]
20. Du, J.; Wu, F.; Wang, J. Intensification of hydrodynamics and heat transfer characteristics of coal-char-gas flow in a high solids-flux downer with swirling blade nozzle. *Energy* **2024**, *294*, 130945. [CrossRef]
21. Wang, Z.; Zhong, W.; Yuan, Y.; Cao, X.; Yuan, Y. Heat Transfer Characteristics in Vertical Tube Condensers: Experimental Investigation and Numerical Validation. *Appl. Therm. Eng.* **2024**, *242*, 122435. [CrossRef]
22. Li, Q.; Luo, K.H.; Kang, Q.J.; He, Y.L.; Chen, Q.; Liu, Q. Lattice Boltzmann Methods for Multiphase Flow and Phase-Change Heat Transfer. *Prog. Energy Combust. Sci.* **2016**, *52*, 62–105. [CrossRef]

applied sciences

MDPI

Article

A Numerical Analysis of the Influence of Nozzle Geometric Structure on Spontaneous Steam Condensation and Irreversibility in the Steam Ejector Nozzle

He Li [1], Xiaodong Wang [1,*], Hailong Huang [1], Jiuxin Ning [1] and Jiyuan Tu [2]

[1] School of Mechanical Engineering and Automation, Northeastern University, Shenyang 110819, China; jbjh1510111@stumail.neu.edu.cn (H.L.); neuhhl@stumail.neu.edu.cn (H.H.); jxning@stumail.neu.edu.cn (J.N.)
[2] School of Engineering, RMIT University, Mill Park, VIC 3083, Australia; jiyuan.tu@rmit.edu.au
* Correspondence: xdwang@mail.neu.edu.cn; Tel.: +86-24-8368-7618; Fax: +86-24-8368-0459

Abstract: The spontaneous condensation of wet steam often occurs in the steam ejector nozzle, this deteriorates the performance of the steam ejector. In this paper, we take changing the geometric parameters of the nozzle as the focus of our research and construct an internal connection between steam's condensation behavior and the nozzle's throat radius, the nozzle's divergent section expansion angle, and the nozzle's divergent section length. Our numerical simulation results indicate that an increase in the throat diameter and reduction of the divergent section's expansion angle can inhibit steam condensation behavior, to a certain extent. In particular, the steam condensation behavior will disappear at a 0° expansion angle, but it is not affected by the change in the divergent section's length. In addition, the irreversibility that is seen under different changes to the nozzle's structure parameters was investigated and the results show that the entropy generation that is caused by a phase change accounts for a much higher proportion of the total entropy generation than heat transport and viscous dissipation do. This indicates that steam's condensation behavior makes a large amount of irreversible energy, resulting in energy waste and reducing the performance of the nozzle. Therefore, this study can provide a theoretical reference for suppressing the spontaneous condensation behavior of steam by changing the nozzle's geometry.

Keywords: steam ejector nozzle; nozzle geometric structure; spontaneous condensation of steam; entropy generation; CFD simulation

Citation: Li, H.; Wang, X.; Huang, H.; Ning, J.; Tu, J. A Numerical Analysis of the Influence of Nozzle Geometric Structure on Spontaneous Steam Condensation and Irreversibility in the Steam Ejector Nozzle. *Appl. Sci.* **2021**, *11*, 11954. https://doi.org/10.3390/app112411954

Academic Editor: Liangxing Li

Received: 10 November 2021
Accepted: 13 December 2021
Published: 15 December 2021

1. Introduction

The steam ejector is a simple energy conversion device that is widely used in many industrial fields, such as power generation [1], refrigeration [2], the desalination of seawater [3], fuel cell applications [4], and steam generation systems [5]. As depicted in Figure 1, the steam ejector can convert a higher-pressure primary fluid and a lower-pressure secondary fluid into a mixed fluid of higher-pressure energy and then realize the vacuum effect in the mixing chamber. The nozzle of the steam ejector plays a key role in the above-mentioned process. Its function is to convert the higher-pressure energy of the primary fluid into greater kinetic energy, thereby providing a source of power for the operation of the entire steam ejector system. Therefore, the nozzle directly affects the efficiency of the entire steam ejector system.

The nozzle's geometric structure has a great influence on the performance of the steam ejector and this has been confirmed by many researchers. Six nozzles with different throat diameters were used in a 1 kW steam ejector cooling capacity experimental system that was constructed by Ruangtrakoon et al. [6,7] and it was found that the geometries of the primary nozzle had strong effects on the steam ejector's performance. Yang et al. [8] investigated the effects of nozzles with conical, oval, square, rectangular, and cross-shaped exit sections on a steam ejector's performance by using the CFD method. They discovered that the steam ejector with the nozzle that had a cross-shaped exit section could achieve the largest entrainment ratio. Fu et al. [9] numerically investigated the effects of the primary nozzle's outlet diameter on the performance of steam ejectors, the CFD results illustrated that the entrainment ratio initially increased sharply with the growth of the nozzle's outlet diameter and then it would reduce when the diameter exceeded a certain value. The influence of the throat area ratio between the nozzle and the ejector on the efficiencies in the MED-TVC desalination system was also investigated by Liu et al. [3] and they noted that the mixing efficiency between the primary and secondary fluids and the entrainment ratio of the steam ejector could be improved by adjusting the throat area ratio between the nozzle and the ejector. Other key geometric parameters such as the nozzle's exit position and the nozzle's outlet diameter were optimized by Wu et al. [10] by using the CFD method. This research revealed that the nozzle outlet diameter was the most sensitive influencing factor for the performance of the steam ejector.

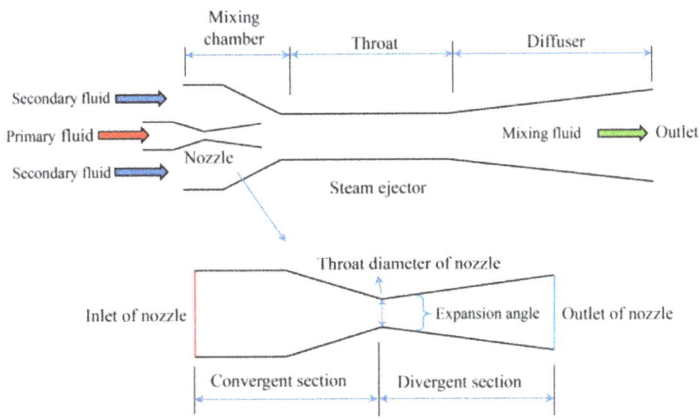

Figure 1. Schematic diagram of steam ejector and nozzle structure.

Apart from the nozzle's geometric structure, the spontaneous condensation behavior of the steam that is in the nozzle also affects the performance of the steam ejector and this is a variable which attracts the attention of many researchers. Moore et al. [11] measured the pressure distribution within the nozzle's condensation zone by conducting a wet-steam tunnel experiment as early as 1973. Subsequently, Moses and Stein [12] documented the homogeneous nucleation and growth of steam droplets using static pressure and laser light scattering methods. Gyarmathy [13] experimentally studied the nucleation characteristics of high-pressure saturated steam and determined the Wilson lines and the droplet size and number count that are present in Laval nozzles. Meanwhile, the influence of the Wilson point position was first theoretically explained in Wilson's early experiments [14]. Bakhtar and Mohammadi Tochai [15] derived a set of numerical algorithms to be used for solving two-dimensional steam two-phase flows in a nozzle. The fundamental equations of gas–droplet multiphase flows were proposed by Young for the calculation of the non-equilibrium condensation of wet steam in nozzles and turbine cascades [16–18].

With the help of a CFD simulation, X. Wang et al. [19] predicted the flow characteristics of a wet steam flow in a nozzle and demonstrated that the steam's condensation behavior could depress the efficiency of the nozzle and reduce the pumping performance of the steam ejector. Moreover, they also studied the effect of primary steam superheating on the steam condensation in a nozzle and the performance of the steam ejector [20]. Sharifi et al. [21,22] numerically studied wet steam flow in the nozzle of a steam ejector and found that the steam's condensation behavior reduced the maximum value of the flow Mach number but improved the compression ratio and entrainment ratio of the steam ejector. In order to improve the accuracy of their numerical calculations, Ding et al. [23] implanted a set of codes based on Young's classical nucleation rate and droplet growth rate into the CFD solver and found that it could finely predict the position of the Wilson point and the wet steam flow properties in a nozzle. Similarly, Abadi et al. developed in-house CFD code that was based on a two-fluid model in order to numerically study the supersonic flow of wet steam with a non-equilibrium phase change in a high-pressure nozzle [24] and high-pressure thermo-compressor [25]. A condensing steam flow mode with droplet size distribution in a Laval nozzle was also implemented into in-house CFD code by Wroblewski and Dykas [26]. A piece of Euler–Euler two-phase flow model code that was optimized by Yang et al. was validated to be able to capture the complicated condensation characteristics of water vapor in the Laval nozzle [27]. These researchers also found that wet steam flow in a nozzle would reduce the performance of the steam ejector more than dry steam. This was reflected in the entrainment ratio that decreased from a score of 0.6973 (which was obtained by the dry steam model) to 0.6242 (which was obtained by the wet steam model) under the given operation condition [28]. Using the multi-objective genetic algorithm method that is based on the modified condensation model, Zhang et al. [29] optimized the primary nozzle of a steam ejector in order to improve its entrainment ratio.

With regard to the aforementioned investigations into steam condensation in the nozzle of a steam ejector, these researchers have only paid attention to the impact of the spontaneous condensation behavior of steam on the inner flow and performance of the steam ejector and the optimization of numerical algorithms in order to improve the accuracy of CFD simulations. However, there are few studies on the influence of a nozzle's geometric structure on its internal steam condensation behavior. Ariafar et al. [30] numerically investigated the wet steam flow in nozzles with area ratios of 11, 18 and 25. They observed that all of the nozzles' static pressures that were obtained by the wet steam model were higher than those for the ideal gas model, while no significant difference existed in the nozzles' momentum. C. Wang et al. [31] also increased the area ratio of the nozzle and found that it reduced the steam condensation's intensity and nozzle's pressure, but increased the liquid mass fraction.

In this article, a wet steam model with entropy generation rates that are relevant to three different mechanisms (viscous dissipation, heat transport, and phase change) was innovatively established in order to numerically study the behavior of wet steam condensation in a nozzle. Apart from the nozzles' throat radii, the effects of the divergent section expansion angle and divergent section length of the nozzle on the spontaneous condensation of steam were investigated. Importantly, the distributions of the entropy generation areas in the nozzles with different geometric structures were visualized for the first time. Finally, the relationships between the nozzle's mass flow rate, total entropy production and the liquid mass fraction of the nozzle for three types of geometric structures were established.

2. Mathematical Models Description

2.1. Governing Equation

The internal steam flow of a nozzle is governed by the compressible steady-state axisymmetric form of the fluid flow conservation equation. The Navier–Stokes equations are more suitable for variable density flows [32]. The total energy equation, including viscous dissipation, was considered and coupled with the gas law. The governing equations of continuity, momentum, and energy can be written as follows:

The continuity equation:

$$\frac{\partial \rho}{\partial t} + \frac{\partial}{\partial x_i}(\rho u_i) = 0 \quad (1)$$

The momentum equation:

$$\frac{\partial}{\partial t}(\rho u_i) + \frac{\partial}{\partial x_j}(\rho u_i u_j) = -\frac{\partial P}{\partial x_i} + \frac{\partial \tau_{ij}}{\partial x_j} \quad (2)$$

The energy equation:

$$\frac{\partial}{\partial t}(\rho E) + \frac{\partial}{\partial x_j}(u_j(\rho E + P)) = -\frac{\partial}{\partial x_j}(\alpha_{eff}\frac{\partial T}{\partial x_j}) + \frac{\partial}{\partial x_j}(u_i\tau_{ij}) \tag{3}$$

$$\rho = \frac{P}{RT} \tag{4}$$

$$\tau_{ij} = \mu_{eff}(\frac{\partial u_i}{\partial x_j} + \frac{\partial u_j}{\partial x_i}) - \frac{2}{3}\mu_{eff}\frac{\partial u_k}{\partial x_k}\delta_{ij} \tag{5}$$

where u_i, u_j, u_k are components of the velocity, τ_{ij} is the stress tensor of the fluid, E is the fluid energy, α_{eff} is the effective thermal conductivity, μ_{eff} is the effective viscosity of the fluid, and δ_{ij} is the unit tensor.

2.2. Turbulence Modeling

The k-ω SST turbulence model was adopted in this study based on its higher prediction accuracy for global and local flow phenomena, which has been confirmed by many researchers [33–36]. It can be described as below [37]:

$$\frac{\partial}{\partial x_i}(\rho k u_i) = \frac{\partial}{\partial x_j}(\Gamma_k\frac{\partial k}{\partial x_j}) + G_k - Y_k + S_k \tag{6}$$

$$\frac{\partial}{\partial x_j}(\rho \omega u_j) = \frac{\partial}{\partial x_j}(\Gamma_\omega\frac{\partial \omega}{\partial x_j}) + G_\omega - Y_\omega + D_\omega + S_\omega \tag{7}$$

where k is the turbulent kinetic energy and ω is the specific dissipation rate. Definitions of the terms G, Y, S, and D can be found in Ref. [37].

2.3. Wet Steam Flow Transport Equation

2.3.1. Liquid Phase Mass Fraction Transport Equation

The transport equation governing the mass fraction of the condensed liquid phase is written as [38]:

$$\frac{\partial(\beta\rho)}{\partial t} + \frac{\partial}{\partial x_i}(\rho\beta u_i) + \frac{\partial}{\partial xj}(\rho\beta u_j) = \Gamma \tag{8}$$

where Γ is the mass generation rate due to condensation and evaporation, which is correlated to the nucleation rate I (the number of new droplets per unit volume per second) and the growth or demise of these droplets [38]:

$$\Gamma = \frac{4}{3}\pi\rho_l I r^{*3} + 4\pi\rho_l\eta\bar{r}^2\frac{\partial\bar{r}}{\partial t} \tag{9}$$

where r^* is the critical droplet radius, above which the droplet will grow and below which the droplet will evaporate. This can be written as [17]:

$$r^* = \frac{2\sigma}{\rho_l RT\ln(s)} \tag{10}$$

In the above expression, s is defined as the ratio of the steam pressure to the saturation pressure corresponding to steam temperature:

$$s = \frac{P}{P_{sat}(T)} \tag{11}$$

2.3.2. Nucleation Rate Equation

In this study, the classical theory of nucleation that is based on the homogeneous nucleation theory was used to calculate the number of liquid droplets; it is written as below [17]:

$$I = \frac{q_c}{(1+\theta)}\left(\frac{\rho_v^2}{\rho_l}\right)\sqrt{\frac{2\sigma}{M^3\pi}}\exp\left(-\frac{4\pi r^{*2}\sigma}{3K_bT}\right) \tag{12}$$

where θ is a non-isothermal correction factor, which is given by [17]:

$$\theta = \frac{2(\gamma-1)}{(\gamma+1)}\left(\frac{h_{lv}}{RT}\right)\left(\frac{h_{lv}}{RT}-0.5\right) \tag{13}$$

where γ is the ratio of the specific heats and it is taken as 1.32.

The wet steam's density can be determined from the vapor density ρ_v and liquid phase mass fraction β. This equation is written as:

$$\rho = \frac{\rho_v}{1-\beta} \tag{14}$$

2.3.3. Droplet Growth Rate

The size of the droplets is affected by two mechanisms: the transfer of mass from the vapor to the droplets and the transfer of heat from the droplets to the vapor in the form of latent heat [38]. This can be written as:

$$\frac{\partial \bar{r}}{\partial t} = \frac{P}{h_{lv}\rho_l\sqrt{2\pi RT}}\frac{\gamma+1}{2\gamma}c_p(T_d-T) \tag{15}$$

2.3.4. Droplet Density Transport Equation

The density of the droplets' transport equation was obtained from Ref. [37] and is described as

$$\frac{\partial(\rho\eta)}{\partial t} + \frac{\partial}{\partial x_i}(\rho\eta u_i) + \frac{\partial}{\partial x_j}(\rho\eta u_j) = \rho I \tag{16}$$

where η is the number of droplets per unit volume. This can be expressed as:

$$\eta = \frac{\beta}{(1-\beta)V_d(\rho_l/\rho_v)} \tag{17}$$

In this equation, V_d is the average droplet volume, which is defined as:

$$V_d = \frac{4}{3}\pi\bar{r}^3 \tag{18}$$

2.3.5. Equation of State

The wet steam equation of state, which relates the pressure to the vapor density and temperature, is given by [16]:

$$P = \rho_v RT(1 + B\rho_v + C\rho_v^2) \tag{19}$$

where B and C are the second and third viral coefficients, respectively, and are functions of temperature.

Besides this equation, the wet steam isobaric specific heat capacity C_p, specific enthalpy h, and specific entropy s are described as below [17], respectively:

$$\begin{aligned} C_p = C_{p0}(T) + R\Big\{&\Big[(1-\alpha_v T)(B - T\tfrac{dB}{dT}) - T^2\tfrac{d^2B}{dT^2})\Big]\rho_v \\ &+\Big[(1-2\alpha_v T)C + \alpha_v T^2\tfrac{dC}{dT} - T^2\tfrac{d^2C}{dT^2}/2\Big]\rho_v^2\Big\} \end{aligned} \tag{20}$$

$$h = h_0(T) + RT\left[(B - T\frac{dB}{dT})\rho_v + (C - T\frac{dC}{dT}/2)\rho_v^2\right] \tag{21}$$

$$s = s_0(T) + \left[R\ln\rho_v + (B + T\frac{dB}{dT})\rho_v + (C + T\frac{dC}{dT}/2)\rho_v^2\right] \tag{22}$$

where C_{p0}, h_0, s_0 are the standard state isobaric specific heat capacity, enthalpy, and entropy, respectively.

2.4. Entropy Generation Rate

In this article, the total entropy generation rate of a wet steam flow is defined as:

$$S_{gen} = S_{gen,\mu} + S_{gen,h} + S_{gen,l} \tag{23}$$

The entropy generation rates of viscous dissipation $S_{gen,\mu}$ and convection heat transfer $S_{gen,h}$ are used and described as the sum of the part of the total amount of entropy that is generated by the mean gradients component and the other part that is generated by the fluctuating component [38]:

$$S_{gen,\mu} = (S_{gen,\mu})_{mean} + (S_{gen,\mu})_{fluc} \tag{24}$$

$$S_{gen,h} = \left(S_{gen,h}\right)_{mean} + \left(S_{gen,h}\right)_{fluc} \tag{25}$$

where $S_{gen,\mu}$ is the entropy generation rate that is produced by viscous dissipation and the $S_{gen,h}$ originated from the heat transfer. The entropy generation rates of the mean gradients are given as:

$$\left(S_{gen,h}\right)_{mean} = \frac{k}{T^2}\left[\left(\frac{\partial T}{\partial x}\right)^2 + \left(\frac{\partial T}{\partial y}\right)^2 + \left(\frac{\partial T}{\partial z}\right)^2\right] \tag{26}$$

$$\begin{aligned}(S_{gen,\mu})_{mean} = \frac{\mu}{T}\Bigg\{2\left[\left(\frac{\partial u_i}{\partial x}\right)^2 + \left(\frac{\partial u_j}{\partial y}\right)^2 + \left(\frac{\partial u_z}{\partial z}\right)^2\right] + \left(\frac{\partial u_i}{\partial y} + \frac{\partial u_j}{\partial x}\right)^2 \\ + \left(\frac{\partial u_i}{\partial z} + \frac{\partial u_z}{\partial x}\right)^2 + \left(\frac{\partial u_j}{\partial z} + \frac{\partial u_z}{\partial y}\right)^2\Bigg\}\end{aligned} \tag{27}$$

and the fluctuating ones are written as:

$$\left(S_{gen,h}\right)_{fluc} = \frac{\alpha^T}{\alpha}\left(S_{gen,h}\right)_{mean} \tag{28}$$

$$(S_{gen,\mu})_{fluc} = \frac{\rho\varepsilon}{T} \tag{29}$$

Besides this, the phase change generation rate can be expressed as [39]:

$$S_{gen,l} = \Gamma h_{lv}(\frac{1}{T_v} - \frac{1}{T_l}) \tag{30}$$

3. Numerical Modeling

3.1. The Dimensions of the Nozzle and Numerical Solution Settings

In this study, the nozzle B from the works of Moore et al. [11] was selected as the original structure (see Table 1) that was used in order to investigate the effects of the throat radius, expansion angle of the divergent section, and length of the divergent section on the steam's spontaneous condensation. The structures of the nozzles with different throat radii, expansion angles, and divergent section lengths are shown in Figure 2; all the geometric parameters are summarized in Table 1.

Table 1. Geometric structure of the nozzle B.

	Coordinate Value (m)	Point A	Point B	Point C	Point D
Throat	x	−0.25	−0.2	0	0.5
Radius	y	0.0563	0.0563	0.025/0.03/0.035/	0.072
				0.04/0.045/0.05	
Expansion	x	−0.25	−0.2	0	0.5
Angle	y	0.0563	0.0563	0.05	0.056/0.063/0.072
					0.081/0.089/0.98
Divergent	x	−0.25	−0.2	0	0.2~0.5
Section Length	y	0.0563	0.0563	0.5	0.059/0.063/0.068
					0.072/0.076/0.081
Nozzle B	x	−0.25	−0.2	0	0.5
Structure	y	0.0563	0.0563	0.05	0.072

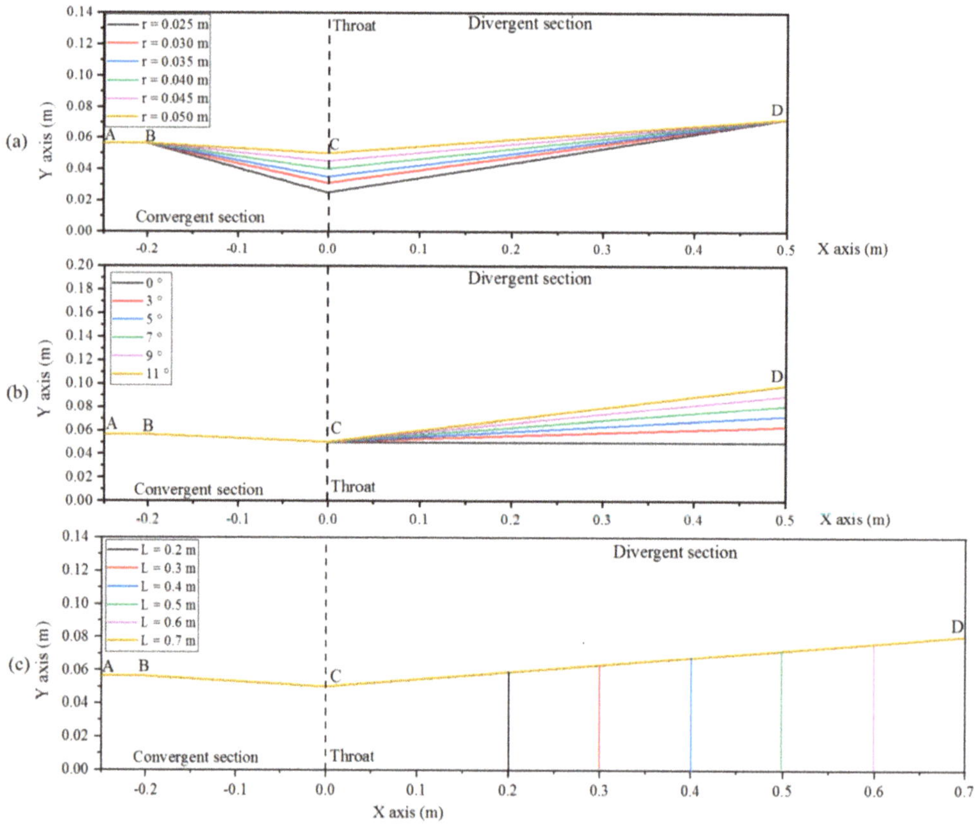

Figure 2. Geometric structure of the nozzle B with: (**a**) different throat radii, (**b**) different expansion angles of the divergent section, and (**c**) different lengths of the divergent section.

In terms of the numerical solution settings, the commercial software Ansys Fluent 15.0, with an implicit density-based solver, was used to solve the flow-governing equations and the wet steam model was selected. The conservation Equations (1)–(5) for the vapor phase were directly solved in the Fluent solver, while the governing Equations (6)–(22) for the liquid phase and the phase change were performed when the wet steam model was called in the solver. The Equations (23)–(30) were implanted into the Fluent solver by UDS and UDF functions and then the entropy generation rate and entropy production rate (the volume integrals for the entropy generation rate) could be solved in every cell of the computational domain and displayed in the post-processing stage. The k-ω SST turbulence model was used. This model did not need the wall function to be set based on its special algorithm [32].

To save computing costs and time, a 2D axisymmetric geometry was adopted instead of a 3D geometry. It has been proven that there is no obvious difference between the two calculation results [40]. The nozzle inlet and outlet were set to the pressure inlet and outlet, respectively. The wall was set to adiabatic and no-slip. All the initial boundary conditions, boundary condition settings, and properties of the working fluid are listed in Tables 2–4, respectively. The second-order upwind scheme was selected for all of the convection terms and central difference discretization was selected for the diffusion terms. The mass flow rate difference between the inlet and outlet and all of the calculated residuals needed to be less than 10−6 in order to ensure iteration convergence.

Table 2. Initial values of the boundary condition.

	Steam Pressure (Pa)	Steam Temperature (K)	Steam Superheat (K)
Nozzle inlet (A)	25,000	354.6	16.5
Nozzle inlet (B)	25,000	357.6	19.5
Nozzle inlet (C)	25,000	358.6	20.5
Nozzle inlet (A/B/C)	2000	300	/

Table 3. Boundary condition settings.

Item	Boundary Condition
Nozzle Wall	Adiabatic and no-slip
Nozzle inlet	Pressure inlet
Nozzle outlet	Pressure outlet

Table 4. Properties of the working fluid.

Property	Value
Dynamic viscosity	1.34×10^{-5} kg/m·s
Thermal conductivity	0.00261 W/m·K
Specific heat capacity	2014.00 J/kg·K
Molecular weight	18.01534 kg·k/mol

3.2. Grid Independence and Validation of the Numerical Approach

The structured grid of nozzle B is presented in Figure 3. To verify the grid's independence, three grid levels were selected; a coarse level with 10k cells, a medium level with 20k cells, and fine level with 41k cells. As illustrated in Figure 4, the droplet nucleation rates of the three grid levels are depicted under the initial boundary condition of nozzle B in Table 2 with the original structure of nozzle B. This reveals that there is a good level of consistency between the medium and fine grid levels. Considering the computational cost and efficiency, the medium-level grid was adopted. In order to further improve the calculations' accuracy, we refined the medium-level mesh in the near-wall area and the region with large pressure gradient variation, as shown in the partially enlarged views in Figure 3. Finally, the grids with 26,784 cells were used to complete the following numerical simulation. Simultaneously, the wall Y plus value was also controlled at around 1 (see Figure 3).

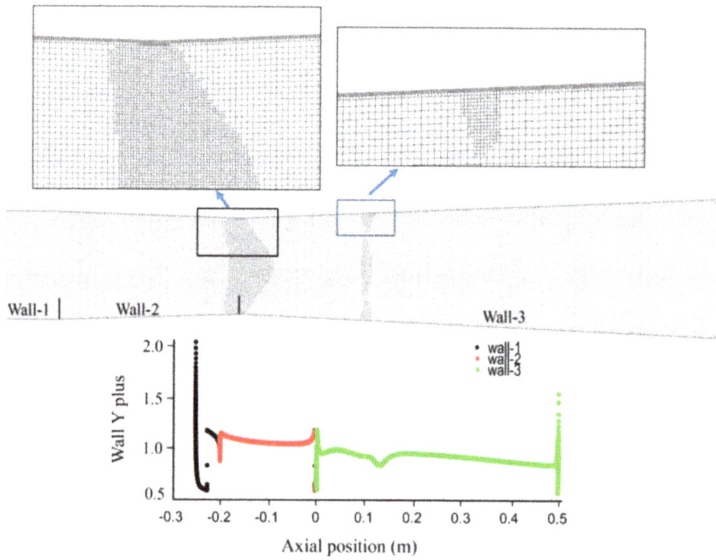

Figure 3. The mesh structure of nozzle B and the wall Y plus value.

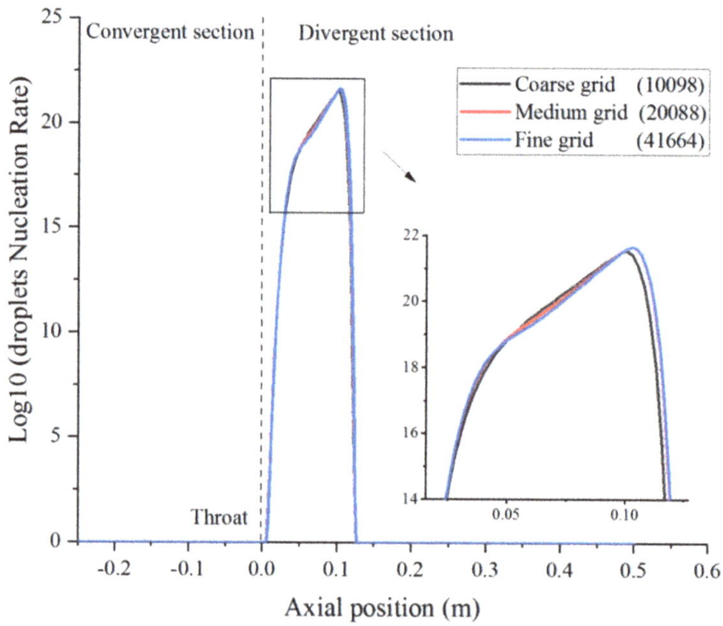

Figure 4. Comparison of the liquid mass fraction under different grid levels.

To identify the simulation's accuracy, the axial static pressure distributions of three nozzles from an experiment that was carried about by Moore et al. [11] under their corresponding boundary conditions (see Table 2) and geometric structures (see Figure 5d) were used, as illustrated in Figure 5. It was found that the CFD results are in close agreement with the present experiment's data. In particular, the wet steam model can accurately predict the onset position and the peak of the condensation shock. Moreover, the average relative error between the simulation result and the experiment value was also calculated and is shown in Figure 5. The average relative error can be obtained by Equation (31):

$$\text{Average relative error } (\%) = \frac{1}{n}\sum_{n=1}^{n}((Expvalue - CFDresult)/Expvalue) \quad (31)$$

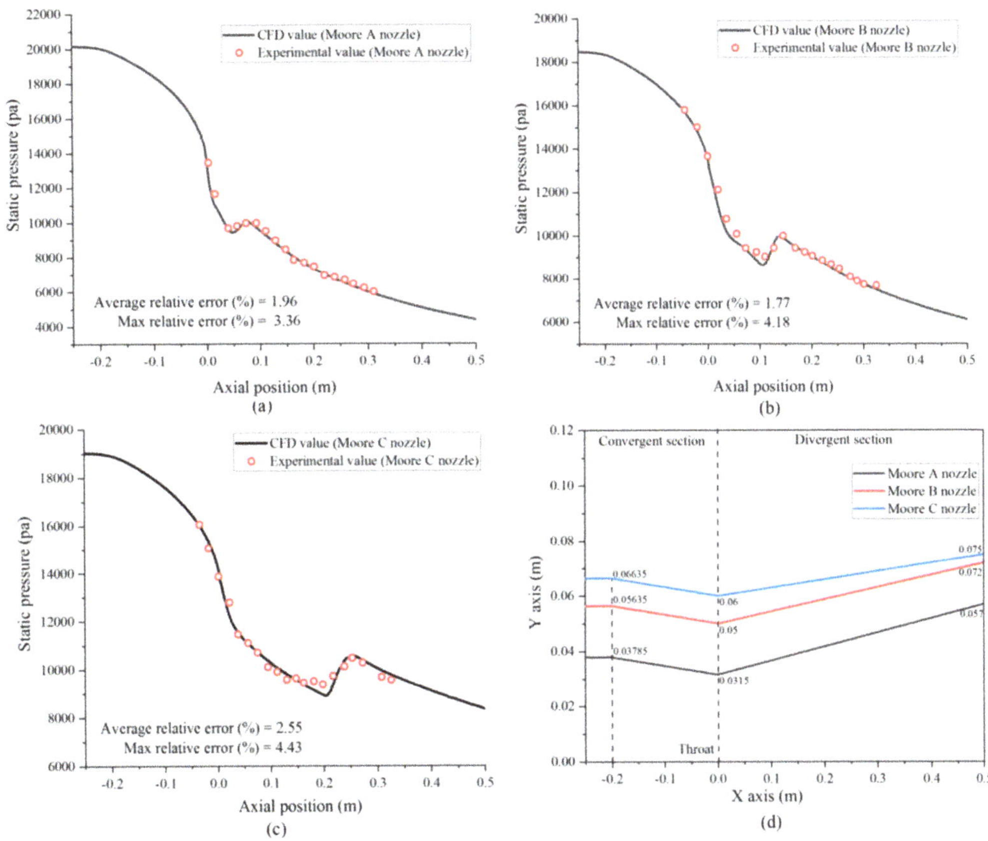

Figure 5. Validation of numerical simulation: (**a**) Moore A nozzle, (**b**) Moore B nozzle, (**c**) Moore C nozzle, and (**d**) the structures of three nozzles.

Among these three nozzles, the biggest average relative error (2.55) and maximum relative error (4.43) were both generated in nozzle C (see Figure 5c). While the smallest average relative error and maximum relative error were 1.77 in nozzle B (see Figure 5b) and 3.96 in nozzle A (see Figure 5a), respectively. This shows that these errors are acceptable and that the wet steam model can fully predict the internal flow process and illustrate the steam condensation phenomenon in a nozzle.

4. Results and Discussion

4.1. The Influence of Nozzle Geometry on Internal Steam Condensation

4.1.1. Nozzle Throat Radius

As steam flows through a nozzle, it expands rapidly and is accompanied by spontaneous condensation. In the flow process of wet steam's spontaneous condensation, the free energy barrier of the molecules does not cause the steam molecules to aggregate and condense immediately, which results in the steam continuing to expand beyond the saturation line and deviating from the equilibrium state. Take a nozzle with the throat radius of 0.05 m as an example, as the steam flows from the convergent section to the divergent section, the temperature falls rapidly in the convergent section but a significant temperature step occurs upstream of the divergent section (see Figure 6a). This can be attributed to the release of the latent heat that has been generated by the steam condensing. At the same time, this latent heat heats the surrounding steam and causes an abrupt increase in pressure (Figure 6b) and the Mach number (Figure 6c) at the same position as the temperature step; this is an effect that is known as the condensation shock. The steam saturation ratio that is described by Equation (11) reflects the transition of the steam from the equilibrium state ($Sr = 1$) to the non-equilibrium state ($Sr \neq 1$). As shown in Figure 6d, the saturation ratio remained almost at 1 in the convergent section, while it suddenly increased to the peak value at the position of condensation shock and then rapidly dropped to 1 in the divergent section. As a result of this, massive droplet cores were formed in the condensation shock position when the saturation ratio was big enough to make the steam molecules overcome their intermolecular rejection of coming together. The droplets' nucleation ratio, in Figure 6e, reflects the fact that the massive droplet cores were generated in the condensation shock region and that, after crossing this condensation shock wave, there was no droplet core appearing in the divergent section. After the droplets' nucleation ratio dropped from the peak value of approximately $\log 10^{22.5}$ to zero, the number of droplets per unit volume was kept constant in the divergent section, as illustrated in Figure 6f.

Moreover, the liquid mass fraction started to rise until it reached its maximum at the nozzle outlet in the divergent section, as depicted in Figure 6g. It is worth noting that the curve of the liquid mass fraction can be divided into two parts: the rapid nucleation zone with slope 1 and the droplet growth zone with slope 2, as presented by the schematic diagram in Figure 6g. Obviously, slope 1 is larger than slope 2. In the rapid nucleation zone, the reason for the steepness of slope 1 of the liquid mass fraction curve is that the massive droplets nucleated, which resulted in the liquid mass fraction (Figure 6g) and droplets' average radius (Figure 6h) increasing sharply in this zone. However, in the droplet growth zone, the droplets' nucleation ratio (Figure 6e) was zero and the droplets per unit volume (Figure 6f) also trended to a constant, which led to those steam molecules that were present in a free state condensing gradually on the droplet cores. Since the droplet growth process was much slower than the nucleation process, this made slope 2 of the liquid mass fraction curve smaller than slope 1. It is also not hard to infer that the average droplet radius grows much slower in the droplet growth zone than it does in the rapid nucleation zone, as shown in Figure 6h.

In general, the wet steam condensation phenomenon under other throat radii has similar characteristics to that which is found under the throat radius of 0.05 m, as shown in Figure 6. However, as the throat radius increased from 0.025 to 0.05 m, there were some differences in the following aspects: The peak position of the condensation shock moved further downstream, as illustrated in Figure 6c. The intensity of the condensation shock was also strengthened, which was reflected in the pressure step change (see Figure 6b). In addition, Figure 6d and e show that the saturation ratio (which was larger than 1) and the droplet nucleation ratio (which was larger than 0) were both enlarged along the axis; meanwhile, the peak values of these two parameter curves decreased with the increase of the throat radius. As seen from Figure 6g, the liquid mass fraction in the divergent section under the throat radius of 0.05m was smaller than that of the others, which can contribute to this radius resulting in the least number of droplets per unit volume (see Figure 6f), although an average droplet radius of 0.0175 μm can be acquired under the throat radius of 0.05 m.

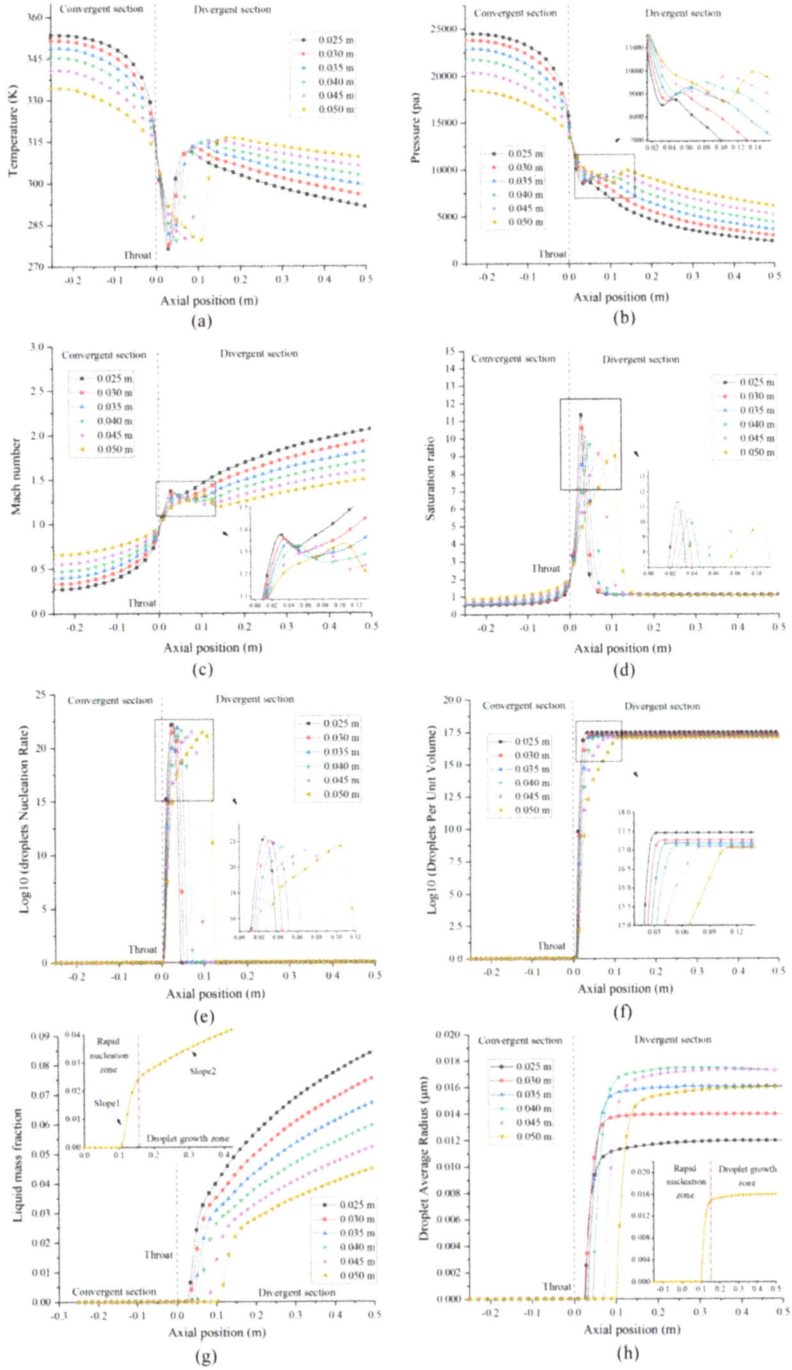

Figure 6. The steam parametric curves along the axis under different nozzle throat radii: (**a**) temperature, (**b**) pressure, (**c**) Mach number, (**d**) saturation ratio, (**e**) droplet nucleation rate, (**f**) droplets per unit volume, (**g**) liquid mass fraction, (**h**) droplet average radius.

4.1.2. Expansion Angle of the Nozzle's Divergent Section

As shown in Figure 7, when the expansion angle of the nozzle's divergent section changes from 0° to 11°, the steam parameters in the nozzle also change to varying degrees. Except for the nozzle with a 0° expansion angle, the steam parameters of the nozzles with other expansion angles had the same changing trends along the axis: continued expansion as the steam passed through the nozzle, a rapid drop in the temperature in the nozzle's convergent section, and the sudden appearance of a temperature step near the nozzle's throat (Figure 7a), which indicates that steam condensation had occurred and a large amount of latent heat was released. The condensation shock that was formed by the steam's condensation resulted in a pressure rise and Mach number reduction in the location of the steam condensation, as illustrated in Figure 7b,c. The droplets' nucleation ratio (Figure 7e) reveals that massive droplet cores were generated, due to the higher saturation ratio (Figure 7d) in the steam condensation zone. Therefore, the formation of these droplet cores sharply increased the liquid mass fraction and the average droplet radius in the rapid nucleation zone, as shown in Figure 7g. Additionally, the volume of droplets per unit remained constant in the divergent section (see Figure 7f), which also caused a slight increase in the average droplet radius (Figure 7h) of the droplets in the droplet growth zone. This is because the steam molecules that were in a free state condensed on the droplet cores, as discussed in Section 4.1.1. In particular, for the nozzle with a 0° expansion angle, the steam hardly condensed at all, resulting in the droplets' nucleation rate, droplets per unit volume, droplets' average radius and liquid mass fraction all equaling zero, as seen in Figure 7.

As shown in Figure 7, we also learned that the enlargement of the nozzle's divergent section's expansion angle reduced the intensity of the condensation shock wave and moved the position of the condensation shock forward (see the enlarged view in Figure 7b,c). The regions of the steam saturation ratio that were larger than 1 and the droplets' nucleation ratio that were larger than 0 were both reduced in accordance with the increase in the throat expansion angle, as depicted in Figure 7d,e. However, the droplets per unit volume (Figure 7f) increased with the enlargement of the expansion angle, which infers that enlarging the expansion angle of the nozzle's divergent section will increase the mass fraction of the liquid phase (Figure 7g), thereby increasing the humidity of the nozzle outlet. As a consequence, we should reasonably reduce the expansion angle of the nozzle's divergent section in order to alleviate the adverse effect of steam's spontaneous condensation in the nozzle.

4.1.3. Length of the Nozzle's Divergent Section

To investigate the effect of a nozzle's divergent section length on steam condensation, we selected six different lengths (0.2, 0.3, 0.4, 0.5, 0.6, and 0.7 m). As can be seen in Figure 8, all of the steam parametric curves were completely overlapped. The only difference was found in those parameter values that were positioned at the nozzle's outlet. Figure 8a demonstrates that the steam condensation released the same amount of latent heat (i.e., the same temperature step) under the different nozzles' divergent section lengths. All of the parametric curves had the same condensation shock position and shock intensity, as illustrated in Figure 8b,c. As the nozzle's divergent section was lengthened, the pressure value at the nozzle's outlet gradually decreased. Compared with the pressure value, the nozzle outlet's Mach number showed the opposite result.

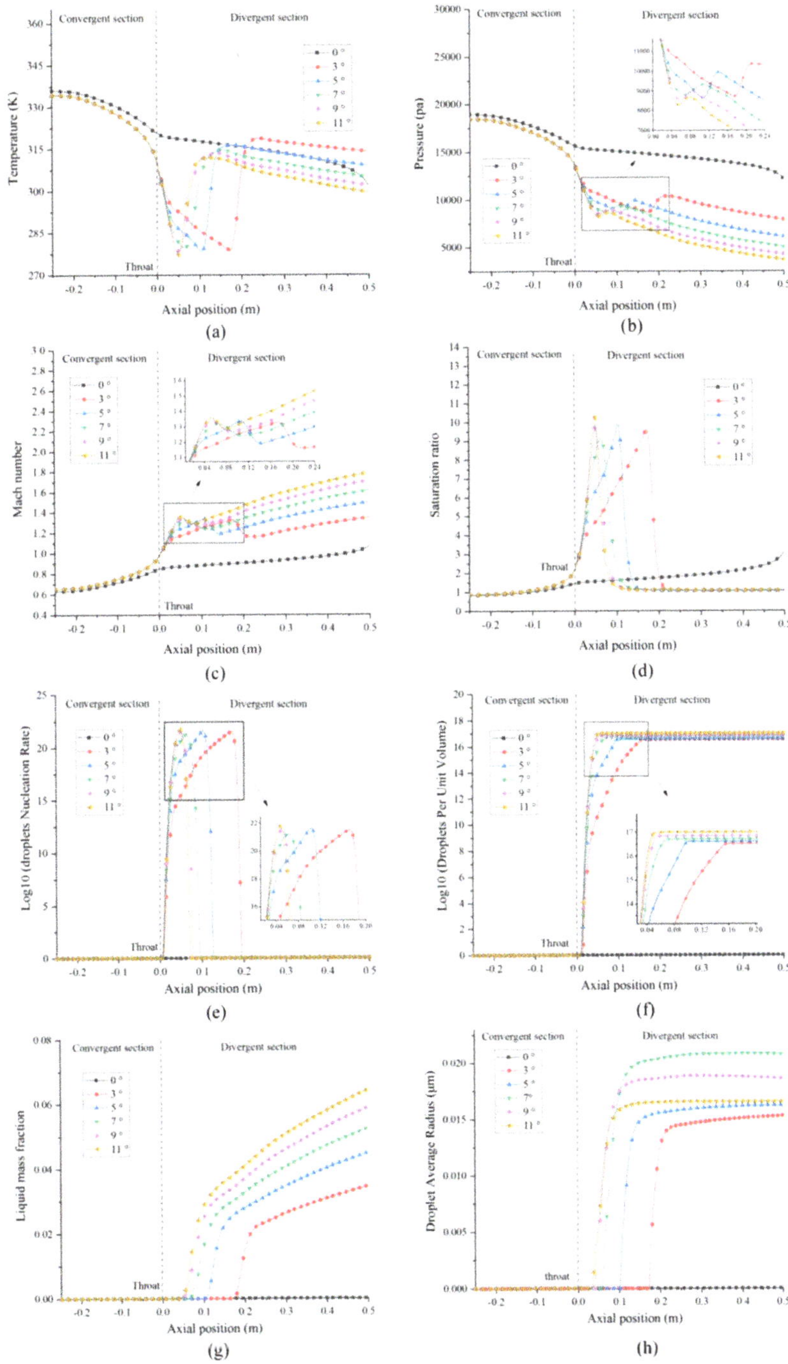

Figure 7. The steam parametric curves along the axis at different expansion angles of the nozzle's divergent section: (**a**) temperature, (**b**) pressure, (**c**) Mach number, (**d**) saturation ratio, (**e**) droplet nucleation rate, (**f**) droplets per unit volume, (**g**) liquid mass fraction, (**h**) droplet average radius.

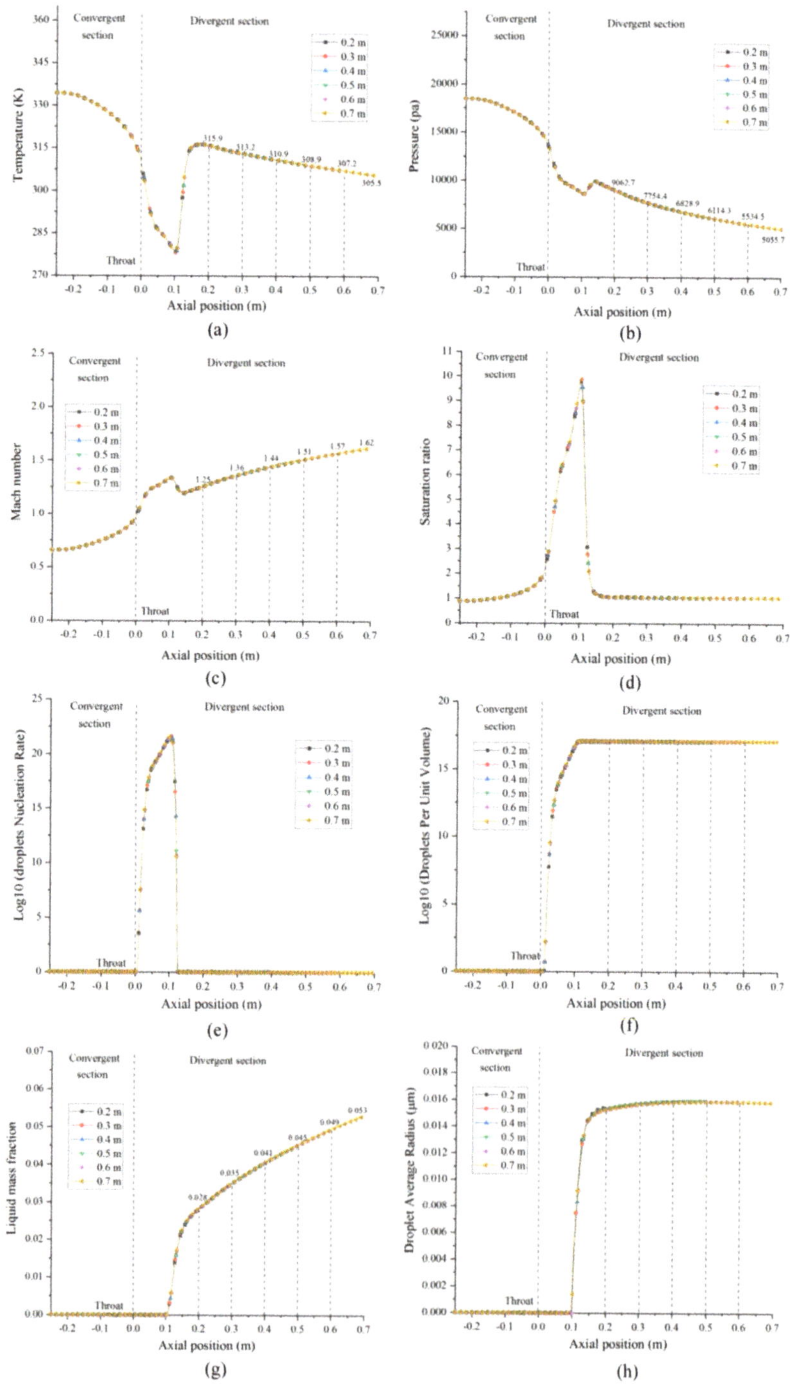

Figure 8. The steam parametric curves along the axis under different lengths of the nozzle's divergent section: (**a**) temperature, (**b**) pressure, (**c**) Mach number, (**d**) saturation ratio, (**e**) droplet nucleation rate, (**f**) droplets per unit volume, (**g**) liquid mass fraction, (**h**) droplet average radius.

Due to the release of the same amount of latent heat from the steam condensation, the same saturation ratio (Figure 8d) and droplets' nucleation ratio (Figure 8e) were formed under the different nozzles' divergent section lengths. The same curves of these two parameters will inevitably lead to the same curve for the droplets per unit volume, as shown in Figure 8f. Figure 8g illustrates that the liquid mass fraction of the nozzle's outlet increased with the increase in the nozzle's divergent section length, which can be attributed to the increase in the droplet number in the nozzle. This is the product of the volume of the nozzle's divergent section and the number of droplets per unit volume.

4.2. Irreversibility Analysis under Different Nozzle Geometries

In this section, the entropy generation rate is investigated in order to identify where the irreversibility occurs, the type of entropy generation, and its effect on the nozzle's performance.

4.2.1. Nozzle Throat Radius

It can be seen from Figure 9 that the total entropy generation rate was mainly generated and concentrated at the nozzle's throat. The area of the entropy generation rate gradually increased with the increase in the throat radius, but the peak value of the total entropy generation rate decreased. This directly caused the total entropy production (the volume integrals for the total entropy generation rate) to decrease from 73.99 W/K under a throat radius of 0.025 m to 71.71 W/K under a throat radius of 0.05 m. Moreover, the position of the total entropy generation rate concentration region gradually moved downstream of the nozzle's divergent section.

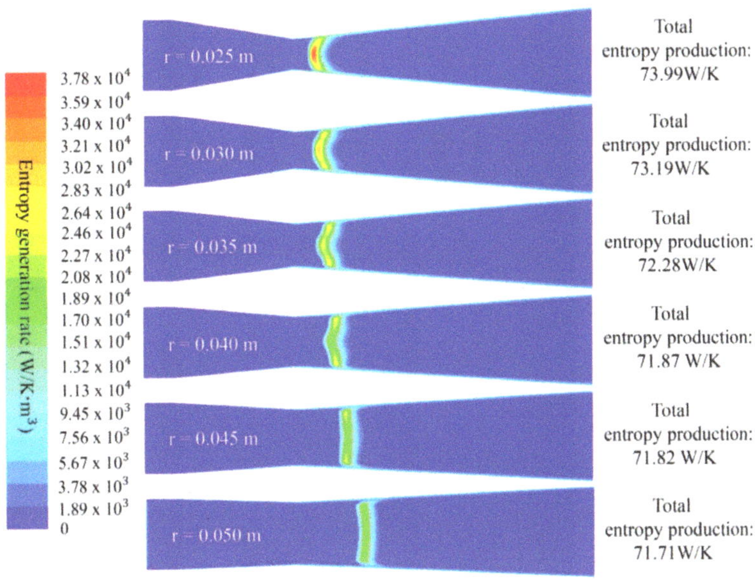

Figure 9. The total entropy generation rate and total entropy production under different nozzle throat radii.

Three types of entropy generation make up the total entropy production: viscous dissipation, heat transport, and phase change entropy generation. These are illustrated in Figure 10 in the form of a stacked column chart. It was found that phase change entropy generation accounted for the largest proportion of entropy, followed by viscous dissipation, and then heat transport. This was found to be true under all of the different nozzle throat radii conditions. The viscous dissipation and the phase change entropy generation were mainly generated in the steam condensation shock region, whereas the heat transport entropy generation was mostly produced at the boundary between the liquid phase and gas phase. As shown in Figure 10, the heat transport and the phase change entropy generation gradually reduced, while the viscous dissipation entropy generation increased with the increase in throat radius. Heat transport and phase change entropy generation reduced from 48.19 and 5.39 W/K to 47.31 and 2.11 W/K, respectively. The viscous dissipation entropy generation increased from 20.41 W/K (the original value) to 22.36 W/K. It is worth noting that the reduction of the phase change entropy generation caused the decrease of the liquid mass fraction and the suppression of the steam's condensation. In addition, it can be more clearly seen from Figure 11 that the liquid mass fraction and the total entropy production both reduced when the nozzle throat radius increased from 0.025 to 0.05 m, but the opposite result was obtained in terms of the mass flow rate. This indicates that properly increasing the nozzle throat radius is beneficial to improving the performance of the nozzle, suppressing the intensity of the spontaneous condensation of steam and reducing energy waste.

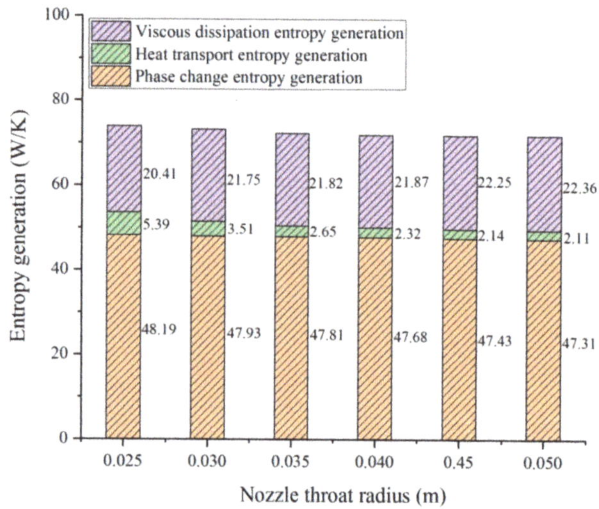

Figure 10. The three types of entropy generation under different nozzle throat radii.

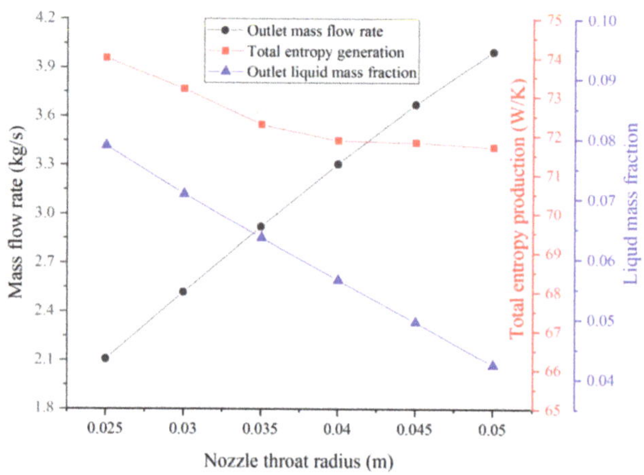

Figure 11. The mass flow rate, total entropy production and liquid mass fraction under different nozzle throat radii.

4.2.2. Expansion Angle of the Nozzle's Divergent Section

As shown in Figure 12, as the nozzle's expansion angle was increased, the concentration region of the total entropy generation rate gradually moved upstream of the nozzle's divergent section and its peak value and area also increased. Therefore, the total entropy production increased from 17.55 W/K at the 0° nozzle expansion angle to 97.97 W/K at the 11° nozzle expansion angle; a nearly five-fold increase. This was due to the condensation behavior of the steam, which almost disappeared at the 0° nozzle expansion angle, so a small total entropy production value was generated.

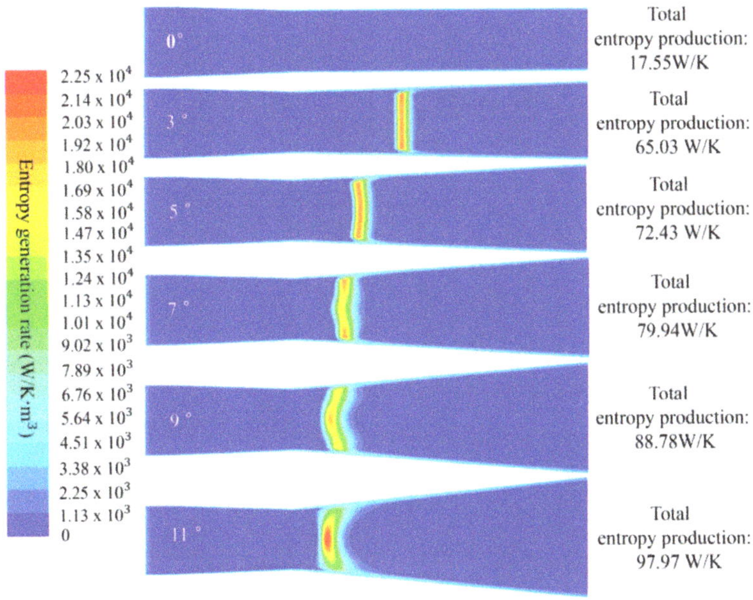

Figure 12. The total entropy generation rate and total entropy production at different expansion angles of the nozzle's divergent section.

Figure 13 illustrates that the enlargement of the nozzle expansion angle changed the composition of the three modes of entropy generation: the nozzle expansion angle changed from 0 to 11°, which resulted in an increase in the viscous dissipation and the heat transport entropy generation from 16.12 and 1.44 W/K to 24.41 and 2.58 W/K, respectively. Moreover, the increase in phase change entropy generation from 0 to 71 W/K also led to an increase in the liquid mass fraction. As already seen for the proportion of entropy production in Section 4.2.1, the largest proportion of entropy production was still generated by phase changes, followed by viscous dissipation and then heat transport.

Figure 14 clearly shows that the curve of the mass flow rate has a steeper slope between the 0° and 3° nozzle expansion angles than that which was found between the 3° and 11° angles. This is because the steam's spontaneous condensation did not happen at the 0° nozzle expansion angle, which led to the generation of smaller values of the total entropy production and liquid mass fraction. Once the expansion angle was larger than 3°, increased entropy production and liquid mass fraction usually led to more energy being converted into useless work and a reduction in the flow velocity of the steam, which directly slowed the growth rate of the mass flow rate, as depicted in Figure 14. Therefore, it is not appropriate to attempt to improve nozzle performance by expanding the expansion angle of the nozzle's divergent section. This caution is evidenced by the lower mass flow rate growth rate, the increase in the total entropy production and the nozzle outlet humidity that were found between the 3° and 11° expansion angles (see Figure 14).

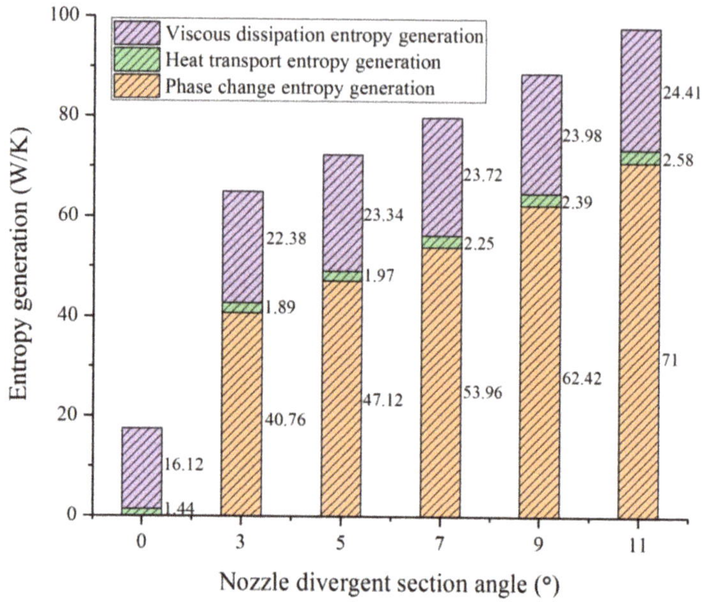

Figure 13. The three types of entropy generation at different expansion angles of the nozzle's divergent section.

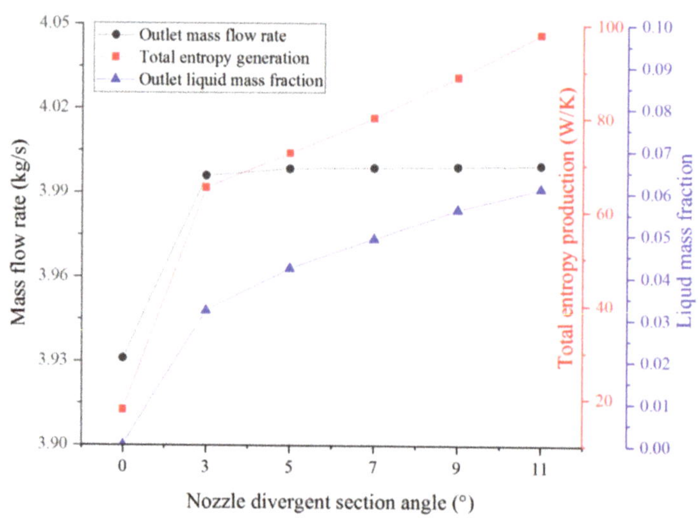

Figure 14. The mass flow rate, total entropy production and liquid mass fraction at different expansion angles of the nozzle's divergent section.

4.2.3. Length of the Nozzle's Divergent Section

The changes in the total entropy generation rate that were caused by the growth of the length of the nozzle's divergent section are presented in Figure 15. It was found that the location and the area of the total entropy generation rate concentration region remained almost the same. The total entropy production increased from 60.04 to 80.59 W/K as the divergent section was lengthened from 0.2 to 0.7 m. Moreover, the three entropy generation processes all increased as the divergent section was lengthened from 0.2 to 0.7 m. Under this condition, the viscous dissipation, heat transport and phase change entropy generation increased from 13.35, 1.41, and 45.27 W/K to 30.06, 2.69, and 47.84 W/K, respectively (see Figure 16). We can see that the phase change entropy generation was still the largest contributor to entropy among the three types of entropy production, under their respective conditions. Interestingly, although there was growth of the total entropy production and liquid mass fraction at the nozzle outlet as the divergent section length increased, we found that the nozzle's performance was almost unaffected. This observation is based on the mass flow rate of the nozzle, which maintained a constant of 3.997 kg/s, as shown in Figure 17.

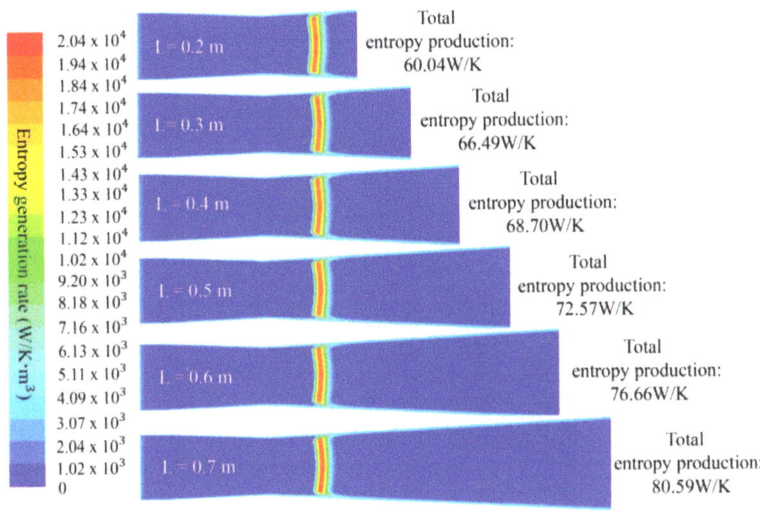

Figure 15. The entropy generation rate and total entropy production under different lengths of the nozzle's divergent section.

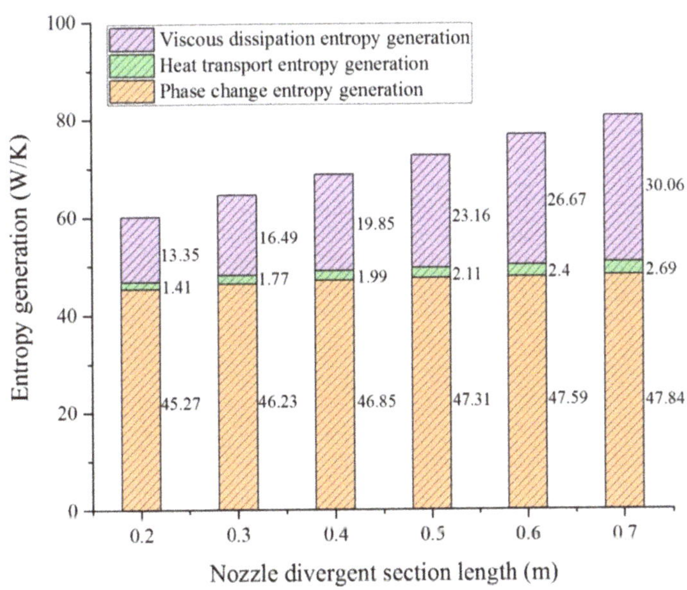

Figure 16. The three types of entropy generation under different lengths of the nozzle's divergent section.

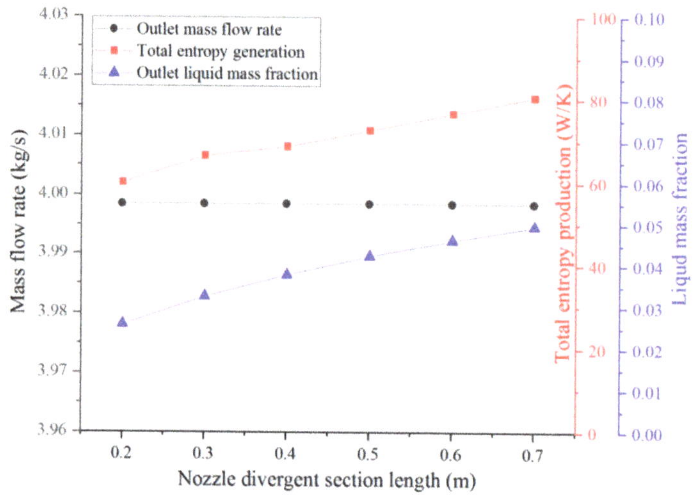

Figure 17. The mass flow rate, total entropy production and liquid mass fraction under different lengths of the nozzle's divergent section.

5. Conclusions

In this paper, the effects of the throat radius, divergent section expansion angle and divergent section length of a nozzle on the spontaneous condensation of steam were numerically investigated. The irreversibility flow in the nozzle was also analyzed by identifying the proportion of entropy generation that could be attributed to three mechanisms (viscous dissipation, heat transport and phase change) and the distributions of the total entropy generation rate under different nozzles' geometric structures. In addition, the nozzle's performance was studied by comparing the mass flow rate, total entropy production and liquid mass fraction. The detailed findings are as follows:

- The increase in the nozzle throat radius enhanced the intensity of the condensation shock and increased the mass flow rate and the pressure at the nozzle outlet. The total entropy production in the nozzle decreased from 11.73 to 10.56 W/K as the throat radius increased from 0.025 to 0.05 m. Therefore, increasing the nozzle throat radius can not only improve the nozzle's performance, but also reduce the waste of energy. In addition, the reduction in liquid mass fraction and phase change entropy generation indicates that the spontaneous condensation behavior of wet steam can be suppressed by increasing the throat radius.
- When the expansion angle of the nozzle's divergent section changed from 0 to 11°, the nozzle's outlet Mach number increased, but the nozzle outlet pressure decreased and the intensity of the condensation shock was weakened. Moreover, there was an increase in the liquid mass fraction, which caused an increase in phase change entropy generation. Therefore, reducing the nozzle's expansion angle inhibited the wet steam's spontaneous condensation behavior, to a certain extent. In particular, the steam's spontaneous condensation behavior disappeared at an expansion angle of 0°.
- Lengthening the nozzle's divergent section did not change the condensation shock's intensity and its position in the nozzle, which resulted in a complete overlap of the curves of the liquid mass fraction, saturation ratio, droplets' nucleation rate and droplets per unit volume in the steam condensation shock region. As the divergent section's length was increased from 0.2 to 0.7 m, the three identified types of entropy generation, total entropy production, and liquid mass fraction all increased, while the mass flow rate remained constant at 3.997 kg/s. Consequently, lengthening the nozzle's divergent section did not improve the nozzle's performance, but it did cause an increase in the humidity in the nozzle's outlet and the wastage of energy.

- No matter how the nozzle's geometric structure changed, the largest proportion of entropy generation was still accounted for by phase changes, followed by viscous dissipation and then heat transport. This indicates that steam's condensation behavior creates a large amount of irreversible energy, resulting in energy waste and reducing the nozzle's performance. Consequently, the spontaneous condensation behavior of wet steam should be suppressed in order to alleviate the performance deterioration that is caused by this behavior when a steam ejector is running.

Author Contributions: Conceptualization, H.L. and X.W.; methodology, H.L. and J.N.; software, H.L and H.H.; validation, J.N. and H.H.; formal analysis, H.L. and J.N.; investigation, H.L., J.N.; resources, X.W.; data curation, H.L. and H.H.; writing—original draft preparation, H.L.; writing—review and editing, H.L. and H.H.; visualization, H.L.; supervision, X.W.; project administration, X.W. and J.T.; funding acquisition, X.W. and J.T. All authors have read and agreed to the published version of the manuscript.

Funding: This research was funded by the National Natural Science Foundation of China (grant no. 51775098) and the Australian Research Council (DP160101953).

Institutional Review Board Statement: Not applicable.

Informed Consent Statement: Not applicable.

Conflicts of Interest: The authors declare no conflict of interest.

Nomenclature

ρ, ρ_v, ρ_l	mixing fluid density, vapor density, and liquid phase density (kg/m^3)
R	gas law constant
T, T_v, T_l	static temperature, vapor temperature, and liquid phase temperature (K)
P	pressure, (Pa)
u_i, u_j, u_k x, y, z	components of velocity, (m/s)
τij	stress tensor
α_{eff}	effective thermal conductivity
μ_{eff}	effective viscosity
δ_{ij}	unit tensor
E	total fluid energy
k	turbulent kinetic energy
ω	specific dissipation rate
μ	dynamic viscosity, (N·s/m^2)
μ_t	eddy viscosity, (N·s/m^2)
β	liquid phase mass fraction
Γ	mass generation rate, (kg/m^{3}s)
S	super saturation ratio
I	number of liquid droplets
η	droplet number density, (1/m^3)
$\bar{r}$	average droplet radius, (m)
r	droplet radius,(m)
r*	critical droplet radius, (m)
KB	Boltzmann constant
σ	droplet surface tension
M	molecular mass, (kg)
θ	non-isothermal correction factor
h	specific enthalpy
h_{lv}	latent heat due to condensation
q_c	evaporation coefficient
γ	ratio of specific heats
Vd	average droplet volume, (m^3)
B, C	virial coefficients, (m^3/kg, m^6/kg^2)
Cp	isobaric heat capacity, (J/(kg·K))

s	specific entropy
Sgen	total entropy generation rate, $(W \cdot K^{-1} m^{-1})$
Sgen,μ	entropy produced by viscous dissipation, $(W \cdot K^{-1} m^{-1})$
Sgen,h	entropy produced by heat transfer, $(W \cdot K^{-1} m^{-1})$
α_t	turbulent thermal diffusivity rate

References

1. Strušnik, D.; Marčič, M.; Golob, M.; Hribernik, A.; Živić, M.; Avsec, J. Energy efficiency analysis of steam ejector and electric vacuum pump for a turbine condenser air extraction system based on supervised machine learning modelling. *Appl. Energy* **2016**, *173*, 386–405. [CrossRef]
2. Thongtip, T.; Aphornratana, S. An experimental analysis of the impact of primary nozzle geometries on the ejector performance used in R141b ejector refrigerator. *Appl. Therm. Eng.* **2017**, *110*, 89–101. [CrossRef]
3. Liu, J.; Wang, L.; Jia, L.; Wang, X. The influence of the area ratio on ejector efficiencies in the MED-TVC desalination system. *Desalination* **2017**, *413*, 168–175. [CrossRef]
4. Brunner, D.A.; Marcks, S.; Bajpai, M.; Prasad, A.K.; Advani, S.G. Design and characterization of an electronically controlled variable flow rate ejector for fuel cell applications. *Int. J. Hydrog. Energy* **2012**, *37*, 4457–4466. [CrossRef]
5. Damiani, L.; Revetria, R. New steam generation system for lead-cooled fast reactors, based on steam re-circulation through ejector. *Appl. Energy* **2015**, *137*, 292–300. [CrossRef]
6. Ruangtrakoon, N.; Aphornratana, S.; Sriveerakul, T. Experimental studies of a steam jet refrigeration cycle: Effect of the primary nozzle geometries to system performance. *Exp. Therm. Fluid Sci.* **2011**, *35*, 676–683. [CrossRef]
7. Ruangtrakoon, N.; Thongtip, T.; Aphornratana, S.; Sriveerakul, T. CFD simulation on the effect of primary nozzle geometries for a steam ejector in refrigeration cycle. *Int. J. Therm. Sci.* **2013**, *63*, 133–145. [CrossRef]
8. Yang, X.; Long, X.; Yao, X. Numerical investigation on the mixing process in a steam ejector with different nozzle structures. *Int. J. Therm. Sci.* **2012**, *56*, 95–106. [CrossRef]
9. Fu, W.; Li, Y.; Liu, Z.; Wu, H.; Wu, T. Numerical study for the influences of primary nozzle on steam ejector performance. *Appl. Therm. Eng.* **2016**, *106*, 1148–1156. [CrossRef]
10. Wu, Y.; Zhao, H.; Zhang, C.; Wang, L.; Han, J. Optimization analysis of structure parameters of steam ejector based on CFD and orthogonal test. *Energy* **2018**, *151*, 79–93. [CrossRef]
11. Moore, P.T.W.; Crane, R.I.; Davidson, B.J. Predicting the fog drop size in wet steam turbines. In *Wet Steam 4 Conference*; Institute of Mechanical Engineers (UK), University of Warwick: Coventry, UK, 1973.
12. Moses, C.A.; Stein, G.D. On the Growth of Steam Droplets Formed in a Laval Nozzle Using Both Static Pressure and Light Scattering Measurements. *J. Fluids Eng.* **1978**, *100*, 311–322. [CrossRef]
13. Gyarmathy, G. Nucleation of steam in high-pressure nozzle experiments. *Proc. Inst. Mech. Eng. Part A J. Power Energy* **2005**, *219*, 511–521. [CrossRef]
14. Gyarmathy, G. *Grundlagen einer Theorie der Nassdampfturbine*; ETH: Zurich, Switzerland, 1962.
15. Bakhtar, F.; Tochai, M.M. An investigation of two-dimensional flows of nucleating and wet steam by the time-marching method. *Int. J. Heat Fluid Flow* **1980**, *2*, 5–18. [CrossRef]
16. Young, J.B. An Equation of State for Steam for Turbomachinery and Other Flow Calculations. *J. Eng. Gas Turbines Power* **1988**, *110*, 1–7. [CrossRef]
17. Young, J.B. Two-Dimensional, Nonequilibrium, Wet-Steam Calculations for Nozzles and Turbine Cascades. *J. Turbomach.* **1992**, *114*, 569–579. [CrossRef]
18. Young, J. The fundamental equations of gas-droplet multiphase flow. *Int. J. Multiph. Flow* **1995**, *21*, 175–191. [CrossRef]
19. Wang, X.-D.; Dong, J.; Wang, T.; Tu, J. Numerical analysis of spontaneously condensing phenomena in nozzle of steam-jet vacuum pump. *Vacuum* **2012**, *86*, 861–866. [CrossRef]
20. Wang, X.; Dong, J.; Li, A.; Lei, H.; Tu, J. Numerical study of primary steam superheating effects on steam ejector flow and its pumping performance. *Energy* **2014**, *78*, 205–211. [CrossRef]
21. Sharifi, N.; Boroomand, M.; Sharifi, M. Numerical assessment of steam nucleation on thermodynamic performance of steam ejectors. *Appl. Therm. Eng.* **2013**, *52*, 449–459. [CrossRef]
22. Sharifi, N.; Boroomand, M.; Kouhikamali, R. Wet steam flow energy analysis within thermo-compressors. *Energy* **2012**, *47*, 609–619. [CrossRef]
23. Ding, H.; Wang, C.; Zhao, Y. An analytical method for Wilson point in nozzle flow with homogeneous nucleating. *Int. J. Heat Mass Transf.* **2014**, *73*, 586–594. [CrossRef]
24. Abadi, S.N.R.; Kouhikamali, R.; Atashkari, K. Two-fluid model for simulation of supersonic flow of wet steam within high-pressure nozzles. *Int. J. Therm. Sci.* **2015**, *96*, 173–182. [CrossRef]
25. Abadi, S.M.A.N.R.; Kouhikamali, R.; Atashkari, K. Non-equilibrium condensation of wet steam flow within high-pressure thermo-compressor. *Appl. Therm. Eng.* **2015**, *81*, 74–82. [CrossRef]
26. Wróblewski, W.; Dykas, S. Two-fluid model with droplet size distribution for condensing steam flows. *Energy* **2016**, *106*, 112–120. [CrossRef]

27. Yang, Y.; Walther, J.H.; Yan, Y.; Wen, C. CFD modeling of condensation process of water vapor in supersonic flows. *Appl. Therm. Eng.* **2017**, *115*, 1357–1362. [CrossRef]
28. Yang, Y.; Zhu, X.; Yan, Y.; Ding, H.; Wen, C. Performance of supersonic steam ejectors considering the nonequilibrium condensation phenomenon for efficient energy utilisation. *Appl. Energy* **2019**, *242*, 157–167. [CrossRef]
29. Zhang, G.; Dykas, S.; Yang, S.; Zhang, X.; Li, H.; Wang, J. Optimization of the primary nozzle based on a modified condensation model in a steam ejector. *Appl. Therm. Eng.* **2020**, *171*, 115090. [CrossRef]
30. Ariafar, K.; Buttsworth, D.; Sharifi, N.; Malpress, R. Ejector primary nozzle steam condensation: Area ratio effects and mixing layer development. *Appl. Therm. Eng.* **2014**, *71*, 519–527. [CrossRef]
31. Wang, C.; Wang, L.; Zou, T.; Zhang, H. Influences of area ratio and surface roughness on homogeneous condensation in ejector primary nozzle. *Energy Convers. Manag.* **2017**, *149*, 168–174. [CrossRef]
32. Ansys, I. *ANSYS-FLUENT 15.0 User's Guide*; ANSYS, Inc.: Pittsburgh, PA, USA, 2013.
33. Besagni, G.; Inzoli, F. Computational fluid-dynamics modeling of supersonic ejectors: Screening of turbulence modeling approaches. *Appl. Therm. Eng.* **2017**, *117*, 122–144. [CrossRef]
34. Varga, S.; Soares, J.; Lima, R.; Oliveira, A. On the selection of a turbulence model for the simulation of steam ejectors using CFD. *Int. J. Low-Carbon Technol.* **2017**, *12*, 233–243. [CrossRef]
35. Mazzelli, F.; Giacomelli, F.; Milazzo, A. CFD modeling of condensing steam ejectors: Comparison with an experimental test-case. *Int. J. Therm. Sci.* **2018**, *127*, 7–18. [CrossRef]
36. Besagni, G.; Cristiani, N.; Croci, L.; Guédon, G.R.; Inzoli, F. Computational fluid-dynamics modelling of supersonic ejectors: Screening of modelling approaches, comprehensive validation and assessment of ejector component efficiencies. *Appl. Therm. Eng.* **2021**, *186*, 116431. [CrossRef]
37. Ishazaki, K.; Ikohagi, T.; Daiguji, H. High-resolution numerical method for transonic non-equilibrium condensation flows through a steam turbine cascade. In Proceedings of the 6th International Symposium on Computational Fluid Dynamics, Lake Tahoe, NV, USA, 4–8 September 1995; pp. 479–484.
38. Herwig, H.; Kock, F. Direct and indirect methods of calculating entropy generation rates in turbulent convective heat transfer problems. *Heat Mass Transf.* **2006**, *43*, 207–215. [CrossRef]
39. Ding, H.; Zhao, Y.; Wen, C.; Wang, C.; Sun, C. Energy efficiency and exergy destruction of supersonic steam ejector based on nonequilibrium condensation model. *Appl. Therm. Eng.* **2021**, *189*, 116704. [CrossRef]
40. Wang, C.; Wang, L.; Zhao, H.; Du, Z.; Ding, Z. Effects of superheated steam on non-equilibrium condensation in ejector primary nozzle. *Int. J. Refrig.* **2016**, *67*, 214–226. [CrossRef]

applied sciences

MDPI

Article

Application of Neural Network and Time-Domain Feature Extraction Techniques for Determining Volumetric Percentages and the Type of Two Phase Flow Regimes Independent of Scale Layer Thickness

Abdullah K. Alanazi [1], Seyed Mehdi Alizadeh [2], Karina Shamilyevna Nurgalieva [3], Slavko Nesic [4], John William Grimaldo Guerrero [5,*], Hala M. Abo-Dief [1], Ehsan Eftekhari-Zadeh [6,*], Ehsan Nazemi [7] and Igor M. Narozhnyy [8]

1 Department of Chemistry, Faculty of Science, Taif University, P.O. Box 11099, Taif 21944, Saudi Arabia; aalanaz4@tu.edu.sa (A.K.A.); h.abodeif@tu.edu.sa (H.M.A.-D.)
2 Petroleum Engineering Department, Australian College of Kuwait, West Mishref 13015, Kuwait; s.alizadeh@ack.edu.kw
3 Department of Development and Operation of Oil and Gas Fields, Saint-Petersburg Mining University, 199106 Saint-Petersburg, Russia; khaibullina_k@mail.ru
4 Faculty of Technology, University of Novi Sad, 21000 Novi Sad, Serbia; slavko.nesic@nis.rs
5 Department of Energy, Universidad de la Costa, Barranquilla 080001, Colombia
6 Institute of Optics and Quantum Electronics, Friedrich-Schiller-University Jena, Max-Wien-Platz 1, 07743 Jena, Germany
7 Imec-Vision Laboratory, Department of Physics, University of Antwerp, 2610 Antwerp, Belgium; ehsan.nazemi@uantwerpen.be
8 Department of Commercialization of Intellectual Activity Resultse Center for Technology Transfer of RUDN University, Mining Oil and Gas Department, RUDN University, 117198 Moscow, Russia; narozhnyy-im@rudn.ru
* Correspondence: jgrimald1@cuc.edu.co (J.W.G.G.); e.eftekharizadeh@uni-jena.de (E.E.-Z.)

Citation: Alanazi, A.K.; Alizadeh, S.M.; Nurgalieva, K.S.; Nesic, S.; Grimaldo Guerrero, J.W.; Abo-Dief, H.M.; Eftekhari-Zadeh, E.; Nazemi, E.; Narozhnyy, I.M. Application of Neural Network and Time-Domain Feature Extraction Techniques for Determining Volumetric Percentages and the Type of Two Phase Flow Regimes Independent of Scale Layer Thickness. *Appl. Sci.* **2022**, *12*, 1336. https://doi.org/10.3390/app12031336

Academic Editor: Liangxing Li

Received: 27 December 2021
Accepted: 24 January 2022
Published: 27 January 2022

Abstract: One of the factors that significantly affects the efficiency of oil and gas industry equipment is the scales formed in the pipelines. In this innovative, non-invasive system, the inclusion of a dual-energy gamma source and two sodium iodide detectors was investigated with the help of artificial intelligence to determine the flow pattern and volume percentage in a two-phase flow by considering the thickness of the scale in the tested pipeline. In the proposed structure, a dual-energy gamma source consisting of barium-133 and cesium-137 isotopes emit photons, one detector recorded transmitted photons and a second detector recorded the scattered photons. After simulating the mentioned structure using Monte Carlo N-Particle (MCNP) code, time characteristics named 4th order moment, kurtosis and skewness were extracted from the recorded data of both the transmission detector (TD) and scattering detector (SD). These characteristics were considered as inputs of the multilayer perceptron (MLP) neural network. Two neural networks that were able to determine volume percentages with high accuracy, as well as classify all flow regimes correctly, were trained.

Keywords: artificial intelligence; feature extraction; scale thickness; two-phase flow; MLP neural network

1. Introduction

In addition to the gamma radiation efficiency technique, which is the basis of this study, there are other techniques, such as hydrostatic, ultrasonic, and hydrometric techniques, that are used to distinguish the flow regime and volume fraction of multiphase flow. The answer to the fundamental question of what the benefit of is implementing a non-invasive method to determine the mentioned parameters can be explained in several parts.

(A) One of the requirements for optimizing the separation process in the oil and gas industry is to have quantitative and sufficient information about the volume fraction of the gas and oil phases.
(B) Understanding the type of flow pattern along with determining the volume fraction of gas and oil phases is a requisite of transfer processes because it is straightforwardly related to a large part of the project economy.
(C) The efficiency of the separation process is highly influenced by the type of flow regime.
(D) Whether the drilling process should continue or stop at any time can only be determined by understanding the volume fraction of each component.

One of the studies conducted years ago to flow metering and determine volume percentage was a study published in 1999 by Abro et al. [1]. Determination of void fraction was done by Abro et al. using low-energy gamma-ray Americium-241 instead of the traditional Caesium-137 source. They tried to determine which detector positions best serve the purpose and placed three detectors at angles of 140-degree, 154-degree, and 180-degree to the source. The EGS4 software package has been chosen as a substrate for volumetric percentage determination with a 3% error for all flow regimes. In 2020, Sattari and his colleagues were able to simulate three common flow regimes in the volumetric percentage of 5–90% using a structure containing a detector and a cesium-137 source [2]. They performed two independent GMDH neural networks to detect volume percentages and flow regimes. In the beginning, they passed the photon spectrum extracted from the detector through a Savitzky–Golay filter to eliminate the existing high-frequency noise, then extracted 7 time-domain characteristics from this spectrum to give them as input of the neural network. This approach could eventually achieve the type of flow regimes and volume percentage prediction with a root mean square error of less than 1.11. Some studies in recent years have shown that gamma radiation can be used as a technique to detect these internal deposits in oil and gas pipelines. Oliviera et al. used a structure with a NaI detector and a source of cesium-137 to inspect the scale deposited pipe in 2015 [3]. In the discussed approach, the detector and the source move simultaneously in 0.5 cm increments. At each stop, the detector records the gamma spectrum passed through the pipe for one minute. The results showed that the presence and thickness of scale in the pipe could be predicted. In reference [4], the researchers simulated three flow regimes in different volumetric percentages and scale thickness inter the pipe; they were able to classify flow regimes with not very high accuracy using SVM neural network and predicted volumetric percentages with an RMSE of less than 3.67 using the MLP neural network. In addition, in a similar study [5], an attempt was made to predict the scale thickness in the pipe independent of the flow regime and volume percentage using the RBF neural network with an RMSE of less than 0.22. Radiation multiphase flowmeters and the applications of artificial intelligence in them can be found in other studies that are available in references [6–15]. Oil and gas pipelines may become internally fouled and have devastating effects. These deposits can reduce the internal diameter, reduce the equipment life cycle, reduce efficiency, and, ultimately, increase costs [16]. The scale deposited in a pipeline is depicted in Figure 1.

Figure 1. An example of a scale formed in a pipeline.

2. Simulation Setup

The modeling of the detection system in this study was performed through version X of Monte Carlo N-Particle Code (MCNPX) [17]. The schematic view of the mentioned detection system is shown in Figure 2. As is apparent, the two-phase flow and the scale formed inside the steel pipe are in the middle, and the dual-energy source and two NaI detectors at a 45-degree angle to each other are on the other side. In this schematic, the stratified flow is considered as an example.

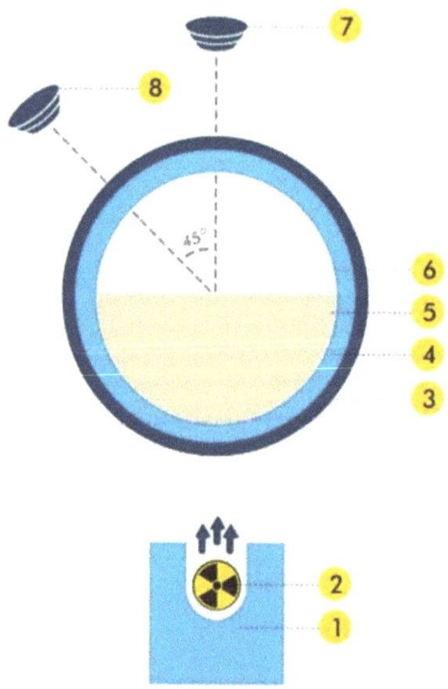

Figure 2. Simulated detection system: 1—shield, 2—dual energy source, 3—steel pipe, 4—scale layer, 5—liquid phase, 6—gas phase, 7—transmission detector, 8—scattering detector.

In the source part, a dual-energy gamma source, including radioisotopes of Cesium-137 and Barium-133 was used, which radiate 0.662 MeV and 0.356 MeV, respectively. In the proposed structure, scattered and transmitted photons are received by two 25.4 mm × 25.4 mm NaI detectors. The detector, which is responsible for recording transmitted photons, was located on the other side of the pipe just opposite the photon transmitter source. The other detector, which is responsible for receiving the scattered photons, was located at a 45-degree angle with the hypothetical line connecting the center of the pipe to the transmission detector. The pipe was considered in this simulation is made of steel with an inner diameter of 20 cm. The scale on the inner wall of the pipe was intended as a circular symmetrical layer of $BaSO_4$ with different thicknesses. Three types of flow regimes with annular, stratified, and homogeneous names were simulated in volume percentages between 10% to 85% with steps of 15% and in seven different scale thicknesses. In our previous work [18], the structure discussed in this article had been validated by several experiments. The obtained detector responses were compared in experimental and simulation. The results showed a good agreement between the simulated results and the experimental data. The uttermost relative difference between the experimental data and simulated data for the detector response was 2.2%. The actual working conditions were dynamic; however, the reference points for flowmeter training were fixed and can be considered static, so experiments and simulations were performed in static conditions. These fixed points were used for flowmeter training to determine the volume fraction and to detect flow regimes in multiphase flowmeters in real conditions. Swift gamma-ray neutron active analysis was considered for quantitative analysis for swift, non-intrusive and online measurements of multiphasic seawater/gas/oil flows in [19]. In this research, all simulations have been considered in static conditions but have been used in real conditions. In [20], conceivable use of transmitted and scattered gamma radiation detection for the characterization of produced water from offshore oil wells has been evaluated. The simulated flow regimes are illustrated in Figure 3. Figure 4 depicts the spectra recorded for three flow regimes with a scale thickness of 2 cm and a void fraction of 55% in both detectors.

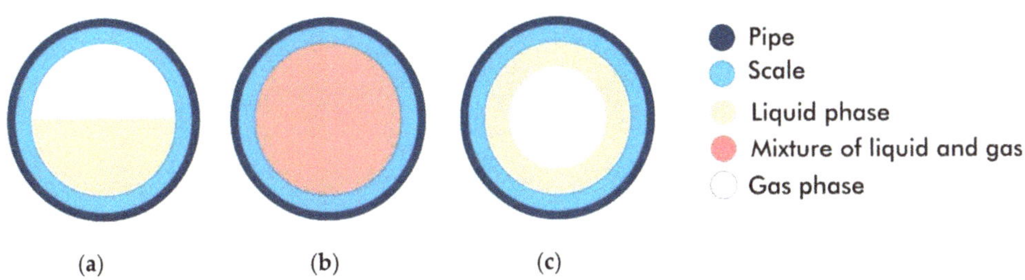

Figure 3. simulated flow regimes (**a**) stratified (**b**) homogeneous (**c**) annular.

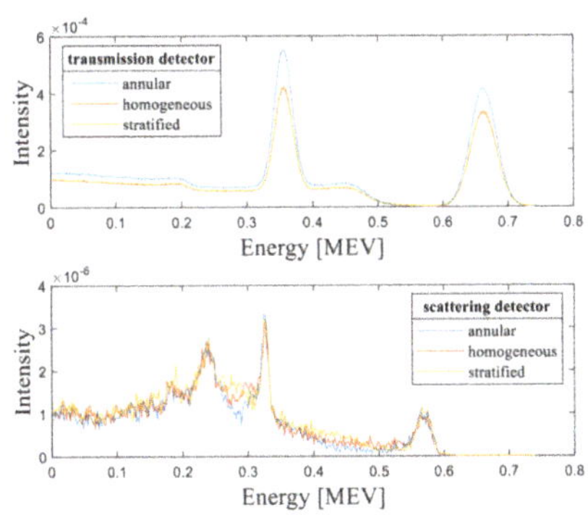

Figure 4. Signal recorded by the transmission and scattering detectors for all three simulated regimes at 2 cm scale thickness and 55% void fraction.

3. Feature Extraction

Feature extraction is a process of defining a set of features that aims to reduce the signal size, but this size reduction should preserve the properties of the signal and make it even more suitable for interpretation. There are several methods for extracting the features of signals. For instance, extraction of characteristics in the frequency domain, extraction of characteristics in the time domain, combined mode of extraction of characteristics in the time-frequency domain, and even other creative methods. Sattari et al. Examined time characteristics that used a creative method to select the characteristics of skewness, kurtosis, and 4th order moment as efficient characteristics [21]. The same creative method was used in the research of Hosseini et al. to select the appropriate frequency characteristics [22]. There are several methods to determine the effective and appropriate properties of the signal, for example, in [2], researchers used the GMDH neural network to determine the effective signal properties. They used the network's self-organizing feature as a tool for feature selection. Feature extraction method using wavelet transform has also been considered by many researchers [23,24]. By extracting the approximate and the details of the received signals, they extract the characteristic features of the signal. Correlation analysis methods have also been studied to determine suitable characteristics in other studies [25]. Nevertheless, the gap in their research was that they did not consider the scale thickness of the inner pipe. In this study, inspired by previous research [21], three time features of skewness, kurtosis, and 4th order moment were extracted as follows:

- 4th order moment:

$$m_4 = \frac{1}{N}\sum_{n=1}^{N}[x(n) - m]^4, \quad m = \frac{1}{N}\sum_{n=1}^{N} x(n) \tag{1}$$

- skewness:

$$g_1 = \frac{m_3}{\sigma^3}, \quad m_3 = \frac{1}{N}\sum_{n=1}^{N}[x_n - m]^3, \ \sigma^2 = \frac{1}{N}\sum_{n=1}^{N}(x_n - m)^2 \tag{2}$$

- kurtosis:

$$g_2 = \frac{m_4}{\sigma^4} \tag{3}$$

4. MLP Neural Network

In the past few decades, various advanced computational approaches, e.g., finite element, numerical linear algebra, statistics, numerical analysis, tensor analysis, and artificial intelligence, have been applied in various fields of study, such as fluid mechanic engineering [26–34], chemical engineering [35–41], electrical and computer engineering [42–72], petrochemical engineering [73–77], petroleum engineering [78–92], mathematics and physics [93–102], and environmental engineering [103–107]. The ANN has been demonstrated to be the most potent technique for classification and prediction among the aforementioned computational methods.

Perceptron is a kind of neural network in terms of computational unit. In fact, perceptron is a single-layer neural network; a multilayer perceptron is called a neural network. Taking of real values of input vectors and calculating a linear composition of these inputs are perceptron's duties. If the result value is less than the threshold value, the perceptron output will be -1 and, otherwise, will be 1. The following equation determines the perceptron output [108,109]:

$$y = f(\sum_{i=1}^{u} w_i x_i + w_i) \tag{4}$$

The page is divided into two parts if the perceptron has two inputs x_1 and x_2 and the equation of the dividing line is defined as follows:

$$w_1 x_1 + w_2 x_2 + w_0 = 0 \tag{5}$$

In the n-dimensional space of instances, perceptron is thought-outed as a hyperplane. Perceptron only learns instances that are linearly separable so that it can completely divide the instances into two parts by a hyperplane and apply the values of -1 to one side and 1 to the other side. Obtaining the values of perceptron weights is the goal of training it so that perceptron produces the real value of training instances. Perceptron is taught according to the following algorithm:

1. Weights get random values
2. For each training instance, perceptron is applied. If the samples are misjudged, the perceptron weight values are corrected.
3. Are all trainings evaluated correctly?
4. Yes, the end of the algorithm.
5. No, back to step 2.

When the network is single-neuron, it loses the ability to implement nonlinear functions. Multilayer perceptron networks (MLP) are very useful and can be offered as a solution because they perform nonlinear mapping with high accuracy, which is considered as the main solution in many engineering problems. MLP is a type of feed-forward network; the output is calculated directly from the input without any feedback. In the MLP network, the neuron model consists of a nonlinear activation function. Nonlinearity, continuity, and derivability at all points are some of the features that an activation function should have. If the activation function is not nonlinear, the network performance will be reduced to the level of monolayer perceptron.

5. Result and Discussion

In this paper, two separate artificial multilayer perceptron neural networks are designed to classify the type of flow regimes and predict the volume percentages independent of the thickness of the scale in the pipe. The inputs of both networks are the characteristics introduced in the previous section- extracted from the recorded signals from both the TD and the SD. In total, 126 simulations were performed in this study, of which about 70% (88 samples), 15% (19 samples), and 15% (19 samples) were used for training, validation, and testing data, respectively. The structure of the classifier network to identify the type of flow regimes is shown in Figure 5. To show the performance of this network, the confusion matrix is plotted for train, validation, and test data in Figure 6. It is very important to say that the output of this network consists of three numbers, 1, 2, and 3, which represent the annular, stratified, and homogeneous regimes, respectively. In addition, the thresholds for the output of this network are defined in such a way that if the output was between 0.5 to 1.5, the output number would be equal to 1, if the output number was between 1.5 to 2.5, the output would return to 2 and if the output was between 2.5 to 3.5, the network declares a homogeneous regime in the output.

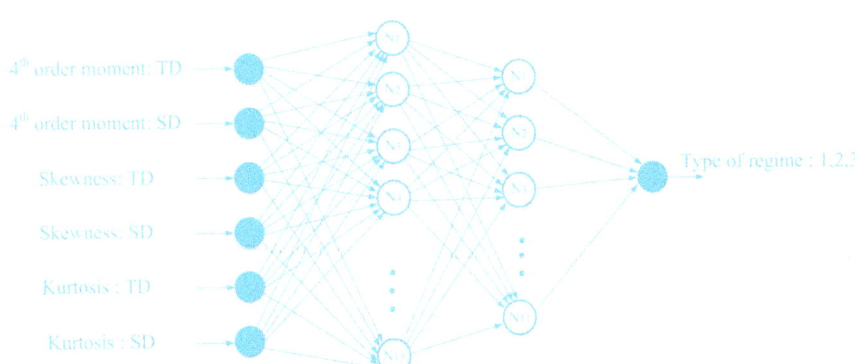

Figure 5. Structure of classifier network.

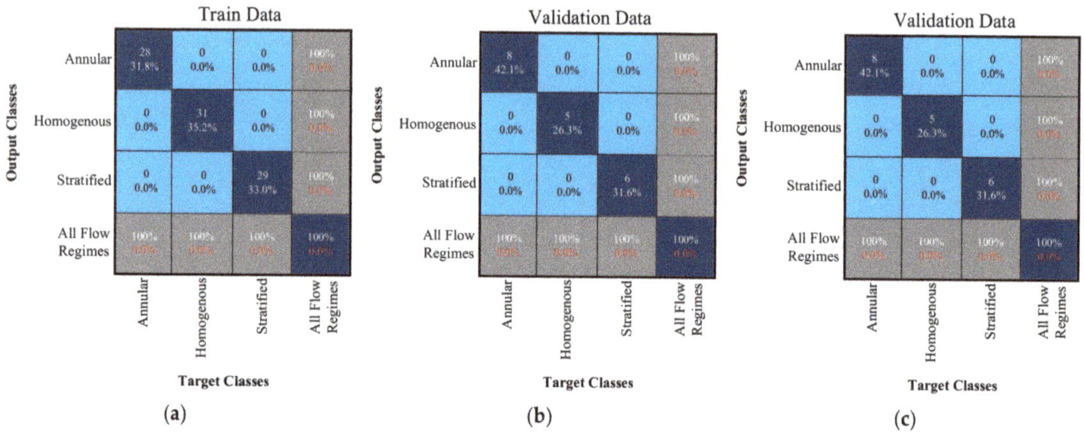

Figure 6. Confusion matrix to show the classification network performance for (**a**) training, (**b**) validation and (**c**) testing data.

Another network has been implemented with the aim of predicting volume percentages; the structure of this network is shown in Figure 7. To obtain the optimal network structure, various networks with one, two, three, and four hidden layers and with the different number of neurons in each layer and different activating functions including linear, log-sigmoid (Logsig), Heaviside, and hyperbolic tangent sigmoid (Tansig) were examined; which is the most optimal structure for both classifier and predictor networks is shown in Table 1. This table includes the number of input neurons, the number of hidden layers, and the number of neurons in each hidden layer, as well as the type of activation function and the number of the epoch. To show the performance of the predictor network, a fitting diagram and the error diagram were used for the three data sets of training, validation, and testing (Figure 8). The fitting diagram shows both the desired output and the output of the designed network on a graph. The more the two diagrams match, the higher the accuracy of the designed network. The error diagram also shows the difference between the desired output and the network output for each of the data. One of the most important parameters in determining the accuracy of predictor networks is determining the accuracy measurement criteria. In this research, criteria named root mean square error (RMSE) and mean relative error (MRE), with the following equations, have been applied. The calculated values for these criteria are listed in Table 2.

$$MRE\% = 100 \times \frac{1}{N}\sum_{j=1}^{N}\left|\frac{X_j(Exp) - X_j(Pred)}{X_j(Pred)}\right| \tag{6}$$

$$RMSE = \left[\frac{\sum_{j=1}^{N}\left(X_j(Exp) - X_j(Pred)\right)^2}{N}\right]^{0.5} \tag{7}$$

where N is the amount of data and '*X (Exp)*' and '*X (Pred)*' stands for the experimental and predicted (ANN) values, respectively.

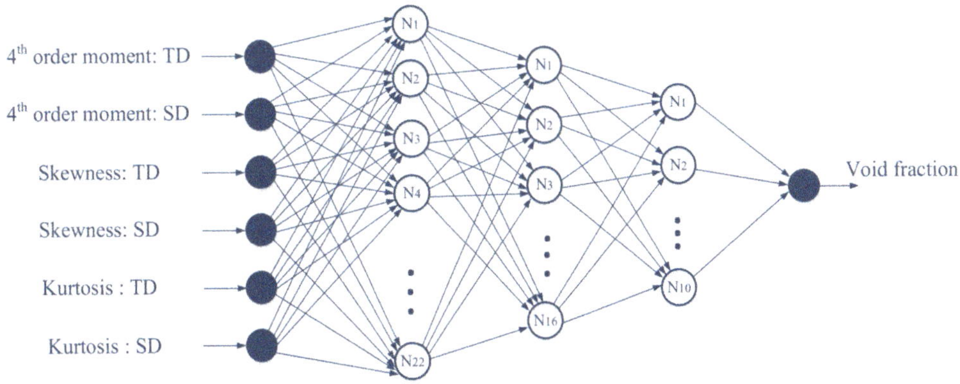

Figure 7. Structure of predictor network.

Table 1. The characteristics of designed networks.

ANN Kind	MLP	
	Classifier	**Predictor**
No. of neurons in input layer	6	6
No. of neurons in the 1st hidden layer	15	22
No. of neurons in the 2nd hidden layer	11	16
No. of neurons in the 3rd hidden layer	-	10
No. of neurons in the output layer	1	1
No. of epoch	680	530
Activation function used for each hidden neuron	Tansig	Tansig

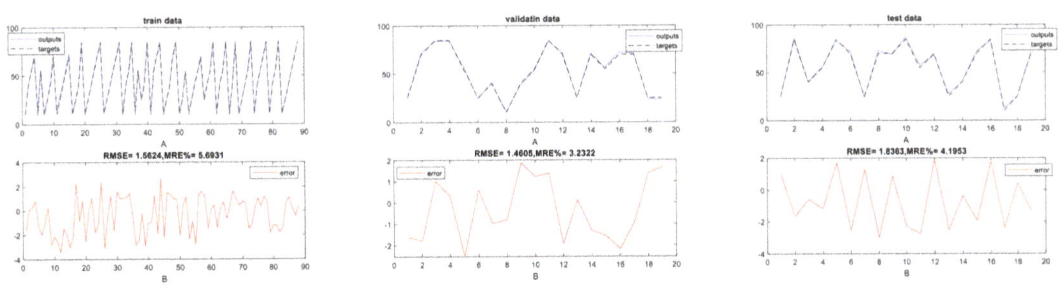

(a) (b) (c)

Figure 8. Fitting and error diagram to show the predictor network performance for (**a**) training, (**b**) validation and (**c**) testing data.

Table 2. Calculated errors for the predictor network.

RMSE	MRE%	Data Set
1.56	5.69	Train
1.46	3.23	Validation
1.83	4.19	Test

6. Conclusions

The present paper, through a creative structure based on radiation, offers an approach for flow pattern and volumetric percentage detection in scale-laden petroleum pipelines. For this purpose, a structure consisting of a dual-energy gamma source, two NaI detectors, and a pipe, to simulate different flow regimes at different volume percentages, as well as to model the scale thickness inside the pipe, was simulated using the Monte Carlo code. After completing all simulations and data collection, to better interpret the collected data, the feature extraction technique in the time-domain was used. The extracted features, which include 4th order moment, skewness, and kurtosis, were considered as neural network inputs. Two designed MLP neural networks were able to fully classify flow regimes and predict volume percentages with an RMSE of less than 1.84, respectively.

Author Contributions: Conceptualization, S.M.A. and I.M.N.; methodology, A.K.A., K.S.N.; software, A.K.A., S.N.; investigation, K.S.N.; data curation, E.N.; writing—original draft preparation, S.M.A., J.W.G.G., E.E.-Z., S.N.; writing—review and editing, H.M.A.-D.; visualization, J.W.G.G.; supervision, E.N.; funding acquisition, I.M.N., A.K.A., E.E.-Z., K.S.N., H.M.A.-D. All authors have read and agreed to the published version of the manuscript.

Funding: We acknowledge support by the Open Access Publication Fund of the Thueringer Universitaets- und Landesbibliothek Jena. This work was supported by the Taif University Researchers Supporting Project grant number (TURSP-2020/266), of Taif University, Taif, Saudi Arabi. This paper has been supported by the RUDN University Strategic Academic Leadership Program. The research was performed at the expense of the subsidy for the state assignment in the field of scientific activity for 2021 No FSRW-2020-0014.

Institutional Review Board Statement: Not applicable.

Informed Consent Statement: Not applicable

Data Availability Statement: Data are contained within the article

Conflicts of Interest: The authors declare no conflict of interest.

References

1. Åbro, E.; Khoryakov, V.A.; Johansen, G.A.; Kocbach, L. Determination of Void Fraction and Flow Regime Using a Neural Network Trained on Simulated Data Based on Gamma-Ray Densitometry. *Meas. Sci. Technol.* **1999**, *10*, 619.
2. Sattari, M.A.; Roshani, G.H.; Hanus, R. Improving the Structure of Two-Phase Flow Meter Using Feature Extraction and GMDH Neural Network. *Radiat. Phys. Chem.* **2020**, *171*, 108725. [CrossRef]
3. Oliveira, D.F.; Nascimento, J.R.; Marinho, C.A.; Lopes, R.T. Gamma Transmission System for Detection of Scale in Oil Exploration Pipelines. *Nucl. Instruments Methods Phys. Res. Sect. A Accel. Spectrometers Detect. Assoc. Equip.* **2015**, *784*, 616–620. [CrossRef]
4. Roshani, M.; Phan, G.T.; AliP, J.M.; Roshani, G.H.; Hanus, R.; Duong, T.; Corniani, E.; Nazemi, E.; KalmounE, M. Evaluation of Flow Pattern Recognition and Void Fraction Measurement in Two Phase Flow Independent of Oil Pipeline's Scale Layer Thickness. *Alex. Eng. J.* **2021**, *60*, 1955–1966. [CrossRef]
5. Alamoudi, M.; Sattari, M.; Balubaid, M.; Eftekhari-Zadeh, E.; Nazemi, E.; Taylan, O.; Kalmoun, E. Application of Gamma Attenuation Technique and Artificial Intelligence to Detect Scale Thickness in Pipelines in Which Two-Phase Flows with Different Flow Regimes and Void Fractions Exist. *Symmetry* **2021**, *13*, 1198. [CrossRef]
6. Roshani, G.H.; Nazemi, E.; Feghhi, S.A.; Setayeshi, S. Flow Regime Identification and Void Fraction Prediction in Two-Phase Flows Based on Gamma Ray Attenuation. *Measurment* **2015**, *62*, 25–32. [CrossRef]
7. Nazemi, E.; Feghhi, S.A.H.; Roshani, G.H.; Peyvandi, R.G.; Setayeshi, S. Precise Void Fraction Measurement in Two-Phase Flows Independent of the Flow Regime Using Gamma-ray Attenuation. *Nucl. Eng. Technol.* **2016**, *48*, 64–71. [CrossRef]
8. Karami, A.; Roshani, G.H.; Salehizadeh, A.; Nazemi, E. The Fuzzy Logic Application in Volume Fractions Prediction of the Annular Three-Phase Flows. *J. Nondestruct. Evaluation* **2017**, *36*, 35. [CrossRef]
9. Roshani, M.; Sattari, M.A.; Ali, P.J.M.; Roshani, G.H.; Nazemi, B.; Corniani, E.; Nazemi, E. Application of GMDH Neural Network Technique to Improve Measuring Precision of a Simplified Photon Attenuation Based Two-Phase Flowmeter. *Flow Meas. Instrum.* **2020**, *75*, 101804. [CrossRef]
10. Sætre, C.; Tjugum, S.-A.; Johansen, G.A. Tomographic Segmentation in Multiphase Flow Measurement. *Radiat. Phys. Chem.* **2014**, *95*, 420–423. [CrossRef]
11. Karami, A.; Roshani, G.H.; Khazaei, A.; Nazemi, E.; Fallahi, M. Investigation of Different Sources in Order to Optimize the Nuclear Metering System of Gas–Oil–Water Annular Flows. *Neural Comput. Appl.* **2018**, *32*, 3619–3631. [CrossRef]

12. Mosorov, V.; Zych, M.; Hanus, R.; Sankowski, D.; Saoud, A. Improvement of Flow Velocity Measurement Algorithms Based on Correlation Function and Twin Plane Electrical Capacitance Tomography. *Sensors* **2020**, *20*, 306. [CrossRef] [PubMed]
13. Salgado, C.; Brandão, L.; Conti, C.; Salgado, W. Density Prediction for Petroleum and Derivatives by Gamma-Ray Attenuation and Artificial Neural Networks. *Appl. Radiat. Isot.* **2016**, *116*, 143–149. [CrossRef]
14. 14. Roshani, M.; Ali, P.J.M.; Roshani, G.H.; Nazemi, B.; Corniani, E.; Phan, N.H.; Tran, H.-N.; Nazemi, E. X-Ray Tube with Artificial Neural Network Model as a Promising Alternative for Radioi-Sotope Source in Radiation Based Two Phase Flowmeters. *Appl. Radiat. Isot.* **2020**, *164*, 109255.
15. Biswal, J.; Pant, H.; Goswami, S.; Samantray, J.; Sharma, V.; Sarma, K. Measurement of Flow Rates of Water in Large Diameter Pipelines Using Radiotracer Dilution Method. *Flow Meas. Instrum.* **2018**, *59*, 194–200. [CrossRef]
16. Candeias, J.; de Oliveira, D.; dos Anjos, M.; Lopes, R. Scale Analysis Using X-Ray Microfluorescence and Computed Radiography. *Radiat. Phys. Chem.* **2014**, *95*, 408–411. [CrossRef]
17. Pelowitz, D.B. *MCNP-X TM User's Manual*; Version 2.5.0. LA-CP-05e0369; Los Alamos National, Laboratory: Los Alamos, NM, USA, 2005.
18. Nazemi, E.; Roshani, G.H.; Feghhi, S.A.H.; Setayeshi, S.; Zadeh, E.E.; Fatehi, A. Optimization of a Method for Identifying the Flow Regime and Measuring Void Fraction in a Broad Beam Gamma-Ray Attenuation Technique. *Int. J. Hydrog. Energy* **2016**, *41*, 7438–7444. [CrossRef]
19. Meric, I.; Johansen, G.A.; Mattingly, J.; Gardner, R. On the Ill-Conditioning of the Multiphase Flow Measurement by Prompt Gamma-Ray Neutron Activation Analysis. *Radiat. Phys. Chem.* **2014**, *95*, 401–404. [CrossRef]
20. Holstad, M.B.; Johansen, G.A. Produced Water Characterization by Dual Modality Gamma-Ray Measurements. *Meas. Sci. Technol.* **2005**, *16*, 1007–1013. [CrossRef]
21. Sattari, M.A.; Roshani, G.H.; Hanus, R.; Nazemi, E. Applicability of time-domain feature extraction methods and artificial intelligence in two-phase flow meters based on gamma-ray absorption technique. *Measurement* **2021**, *168*, 108474. [CrossRef]
22. Hosseini, S.; Roshani, G.; Setayeshi, S. Precise Gamma Based Two-Phase Flow Meter Using Frequency Feature Extraction and only One Detector. *Flow Meas. Instrum.* **2020**, *72*, 101693. [CrossRef]
23. Siavash, H.; Taylan, O.; Abusurrah, M.; Akilan, T.; Nazemi, E.; Eftekhari-Zadeh, E.; Bano, F.; Roshani, G.H. Application of Wavelet Feature Extraction and Artificial Neural Networks for Improving the Performance of Gas–Liquid Two-Phase Flow Meters Used in Oil and Petrochemical Industries. *Polymers* **2021**, *13*, 3647.
24. Balubaid, M.; Sattari, M.A.; Taylan, O.; Bakhsh, A.A.; Nazemi, E. Applications of Discrete Wavelet Transform for Feature Extraction to Increase the Accuracy of Monitoring Systems of Liquid Petroleum Products. *Math.* **2021**, *9*, 3215. [CrossRef]
25. Abdulrahman, B.; Sattari, M.A.; Taylan, O.; Nazemi, E. Application of Feature Extraction and Artificial Intelligence Techniques for Increasing the Accuracy of X-Ray Radiation Based Two Phase Flow Meter. *Mathematics* **2021**, *9*, 1227.
26. Roshani, G.H.; Feghhi, S.A.H.; Mahmoudi-Aznaveh, A.; Nazemi, E.; Adineh-Vand, A. Precise volume fraction prediction in oil-water-gas multiphase flows by means of gamma-ray attenuation and artificial neural networks using one detector. *Measurement* **2014**, *51*, 34–41. [CrossRef]
27. Saberinejad, H.; Keshavarz, A.; Payandehdoost, M.; Azmoodeh, M.R.; Batooei, A. Numerical study of heat transfer performance in a pipe partially filled with non-uniform porous media under Ltne Condition. *Int. J. Num. Methods Heat Fluid Flow* **2018**, *28*, 1845–1865. [CrossRef]
28. Nazemi, E.; Feghhi, S.; Roshani, G.; Setayeshi, S.A.; Peyvandi, R.G. A radiation-based hydrocarbon two-phase flow meter for estimating of phase fraction independent of liquid phase density in stratified regime. *Flow Meas. Instrum.* **2015**, *46*, 25–32. [CrossRef]
29. Roshani, G.H.; Nazemi, E.; Feghhi, S.A.H. Investigation of using 60Co source and one detector for determining the flow regime and void fraction in gas-liquid two-phase flows. *Flow Meas. Instrum.* **2016**, *50*, 73–79. [CrossRef]
30. Rouhi, S.; Xiros, N.; Sadeqi, S.; Ioup, J.; Sultan, C.; VanZwieten, J. CFD validation of the thermodynamic model of a compressed gaseous hydrogen storage tank. Proceeding of 5-6th Thermal and Fluids Engineering Conference (TFEC), Virtual, 26–28 May 2021.
31. Roshani, G.H.; Nazemi, E.; Roshani, M.M. Flow regime independent volume fraction estimation in three-phase flows using dual-energy broad beam technique and artificial neural network. *Neural Comput. Appl.* **2016**, *28*, 1265–1274. [CrossRef]
32. Roshani, G.; Nazemi, E.; Roshani, M. Usage of two transmitted detectors with optimized orientation in order to three phase flow me-tering. *Measurement* **2017**, *100*, 122–130. [CrossRef]
33. Roshani, G.; Nazemi, E. Intelligent densitometry of petroleum products in stratified regime of two phase flows using gamma ray and neural network. *Flow Meas. Instrum.* **2017**, *58*, 6–11. [CrossRef]
34. Roshani, G.; Nazemi, E.; Roshani, M. Intelligent recognition of gas-oil-water three-phase flow regime and determination of volume fraction using radial basis function. *Flow Meas. Instrum.* **2017**, *54*, 39–45. [CrossRef]
35. Du, X.; Tian, W.; Pan, J.; Hui, B.; Sun, J.; Zhang, K.; Xia, Y. Piezo-phototronic effect promoted carrier separation in coaxial P-N junctions for self-powered photodetector. *Nano Energy* **2022**, *92*, 106694. [CrossRef]
36. Karami, A.; Roshani, G.H.; Nazemi, E.; Roshani, S. Enhancing the performance of a dual-energy gamma ray based three-phase flow meter with the help of grey wolf optimization algorithm. *Flow Meas. Instrum.* **2018**, *64*, 164–172. [CrossRef]
37. Roshani, G.H.; Roshani, S.; Nazemi, E.; Roshani, S. Online measuring density of oil products in annular regime of gas-liquid two phase flows. *Measurement* **2018**, *129*, 296–301. [CrossRef]

38. Cai, T.; Dong, M.; Liu, H.; Nojavan, S. Integration of hydrogen storage system and wind generation in power systems under demand response program: A novel P-robust stochastic programming. *Int. J. Hydrogen Energy* **2022**, *47*, 443–458. [CrossRef]
39. Roshani, G.; Hanus, R.; Khazaei, A.; Zych, M.; Nazemi, E.; Mosorov, V. Density and velocity determination for single-phase flow based on radiotracer technique and neural networks. *Flow Meas. Instrum.* **2018**, *61*, 9–14. [CrossRef]
40. Charchi, N.; Li, Y.; Huber, M.; Kwizera, E.A.; Huang, X.; Argyropoulos, C.; Hoang, T. Small mode volume plasmonic film-coupled nanostar resonators. *Nanoscale Adv.* **2020**, *2*, 2397–2403. [CrossRef] [PubMed]
41. Roshani, G.; Nazemi, E.; Roshani, M. Identification of flow regime and estimation of volume fraction independent of liquid phase density in gas-liquid two-phase flow. *Prog. Nucl. Energy* **2017**, *98*, 29–37. [CrossRef]
42. Roshani, S.; Roshani, S. Two-Section Impedance Transformer Design and Modeling for Power Amplifier Applications. *Appl. Comput. Electromagn. Soc. J.* **2017**, *32*, 1042–1047.
43. Lalbakhsh, A.; Alizadeh, S.M.; Ghaderi, A.; Golestanifar, A.; Mohamadzade, B.; Jamshidi, M.B.; Mandal, K.; Mohyuddin, W. A design of a dual-band bandpass ?lter based on modal analysis for modern communication systems. *Electronics* **2020**, *9*, 1770. [CrossRef]
44. Liu, X.; Zheng, W.; Mou, Y.; Li, Y.; Yin, L. Microscopic 3D reconstruction based on point cloud data generated using defocused images. *Measurement and Control* **2021**, *54*, 1309–1318. [CrossRef]
45. Pirasteh, A.; Roshani, S.; Roshani, S. Compact microstrip lowpass filter with ultrasharp response using a square-loaded mod-ified T-shaped resonator. *Turk. J. Electr. Eng. Comput. Sci.* **2018**, *26*, 1736–1746. [CrossRef]
46. Ramtin, A.R.; Nain, P.; Menasche, D.S.; Towsley, D.; de Souza e Silva, E. Fundamental scaling laws of covert ddos attacks. *Performance Evaluation* **2021**, *151*, 102236. [CrossRef]
47. Roshani, S.; Roshani, S. A compact coupler design using meandered line compact microstrip resonant cell (MLCMRC) and bended lines. *Wireless Networks* **2021**, *27*, 677–684. [CrossRef]
48. Ma, Z.; Zheng, W.; Chen, X.; Yin, L. Joint embedding VQA model based on Dynamic Word Vector. *PeerJ Comput. Sci.* **2021**, *7*, e353. [CrossRef] [PubMed]
49. Lalbakhsh, A.; Lotfi Neyestanak, A.A.; Naser-Moghaddasi, M. Microstrip hairpin bandpass filter using modified Minkowski fractal-shape for suppression of second harmonic. *IEICE Trans. Electron.* **2012**, *E95-C*, 378–381. [CrossRef]
50. Pourghebleh, B.; Aghaei Anvigh, A.; Ramtin, A.R.; Mohammadi, B. The importance of nature-inspired meta-heuristic algorithms for solving virtual machine consolidation problem in Cloud Environments. *Cluster Comput.* **2021**, *24*, 2673–2696. [CrossRef]
51. Seyedi, M.; Taher, S.A.; Ganji, B.; Guerrero, J. A hybrid islanding detection method based on the rates of changes in voltage and active power for the Multi-Inverter Systems. *IEEE Trans. Smart Grid* **2021**, *12*, 2800–2811. [CrossRef]
52. Roshani, S.; Roshani, S. Design of a very compact and sharp bandpass diplexer with bended lines for GSM and LTE applica-tions. *AEU. Int. J. Electron. Commun.* **2019**, *99*, 354–360. [CrossRef]
53. Wang, K.; Wang, H.; Li, S. Renewable quantile regression for streaming datasets. *Knowledge-Based Systems* **2022**, *235*, 107675. [CrossRef]
54. Lalbakhsh, A.; Mohamadpour, G.; Roshani, S.; Ami, M.; Roshani, S.; Sayem, A.S.M.; Alibakhshikenari, M.; Koziel, S. Design of a compact planar transmission line for miniaturized rat-race coupler with harmonics suppression. *IEEE Access* **2021**, *9*, 129207–129217. [CrossRef]
55. Roshani, S.; Roshani, S.; Zarinitabar, A. A modified Wilkinson power divider with ultra harmonic suppression using open stubs and lowpass filters. *Analog. Integr. Circuits Signal Process.* **2019**, *98*, 395–399. [CrossRef]
56. Wang, K.; Li, S. Robust distributed modal regression for massive data. *Comput. Stat. Data Anal.* **2021**, *160*, 107225. [CrossRef]
57. Jamshidi, M.B.; Siahkamari, H.; Roshani, S.; Roshani, S. A compact Gysel power divider design using U-shaped and T-shaped reso-nators with harmonics suppression. *Electromagnetics* **2019**, *39*, 491–504. [CrossRef]
58. Jamshidi, M.B.; Roshani, S.; Talla, J.; Roshani, S.; Peroutka, Z. Size reduction and performance improvement of a microstrip Wilkinson power divider using a hybrid design technique. *Sci. Rep.* **2021**, *11*, 7773. [CrossRef] [PubMed]
59. Liu, K.; Ke, F.; Huang, X.; Yu, R.; Lin, F.; Wu, Y.; Ng, D.W. Deepban: A temporal convolution-based communication framework for dynamic WBANS. *IEEE Trans. Commun.* **2021**, *69*, 6675–6690. [CrossRef]
60. Pirasteh, A.; Roshani, S.; Roshani, S. A modified class-F power amplifier with miniaturized harmonic control circuit. *AEU Int. J. Electron. Commun.* **2018**, *97*, 202–209. [CrossRef]
61. Roshani, S.; Roshani, S. Design of a high efficiency class-F power amplifier with large signal and small signal measure-ments. *Measurement* **2020**, *149*, 106991. [CrossRef]
62. Lalbakhsh, A.; Jamshidi, M.; Siahkamari, H.; Ghaderi, A.; Golestanifar, A.; Linhart, R.; Talla, J.; Simorangkir, R.B.; Mandal, K. A compact lowpass ?lter for satellite communication systems based on transfer function analysis. *AEU-Int. J. Electron. Commun.* **2020**, *124*, 153318. [CrossRef]
63. Nazemi, B.; Rafiean, M. Forecasting House Prices in Iran Using GMDH. *Int. J. Hous. Mark. Anal.* **2021**, *14*, 555–568. [CrossRef]
64. Roshani, S.; Roshani, S. Design of a compact LPF and a miniaturized Wilkinson power divider using aperiodic stubs with harmonic suppression for wireless applications. *Wireless Networks* **2019**, *26*, 1493–1501. [CrossRef]
65. Lalbakhsh, A.; Afzal, M.U.; Esselle, K. Simulation-driven particle swarm optimization of spatial phase shifters. In Proceedings of the 2016 International Conference on Electromagnetics in Advanced Applications (ICEAA), Cairns, QLD, Australia, 19–23 September 2016.

66. Hookari, M.; Roshani, S.; Roshani, S. High?efficiency balanced power amplifier using miniaturized harmonics suppressed coupler. *Int. J. RF Microwave Comput.-Aided Eng.* **2020**, *30*, e22252. [CrossRef]
67. Lalbakhsh, A.; Afzal, M.U.; Esselle, K.P.; Smith, S.L. Low-cost nonuniform metallic lattice for rectifying aperture near-field of electromagnetic bandgap resonator antennas. *IEEE Trans. Antennas Propagation* **2020**, *68*, 3328–3335. [CrossRef]
68. Pirasteh, A.; Roshani, S.; Roshani, S. Design of a miniaturized class F power amplifier using capacitor loaded transmission lines. *Frequenz* **2020**, *74*, 145–152. [CrossRef]
69. Lalbakhsh, A.; Afzal, M.U.; Hayat, T.; Esselle, K.P.; Manda, K. All-metal wideband metasurface for near-?eld transformation of medium-to-high gain electromagnetic sources. *Sci. Rep.* **2021**, *11*, 1–9.
70. Lotfi, S.; Roshani, S.; Roshani, S.; Gilan, M.S. Wilkinson power divider with band-pass filtering response and harmonics suppression using open and short stubs. *Frequenz* **2020**, *74*, 169–176. [CrossRef]
71. Lalbakhsh, A.; Ghaderi, A.; Mohyuddin, W.; Simorangkir, R.B.V.B.; Bayat-Makou, N.; Ahmad, M.S.; Lee, G.H.; Kim, K.W. A Compact C-Band Bandpass Filter with an Adjustable Dual-Band Suitable for Satellite Communication Systems. *Electronics* **2020**, *9*, 1088. [CrossRef]
72. Hookari, M.; Roshani, S.; Roshani, S. Design of a low pass filter using rhombus-shaped resonators with an analyticallc equivalent circuit. *Turkish J. Electr. Eng. Comput. Sci.* **2020**, *28*, 865–874. [CrossRef]
73. Yadollahi, A.; Nazemi, E.; Zolfaghari, A.; Ajorloo, A.M. Application of artificial neural network for predicting the optimal mixture of radiation shielding concrete. *Progress Nuclear Energy* **2016**, *89*, 69–77. [CrossRef]
74. Roshani, M.; Phan, G.; Faraj, R.H.; Phan, N.H.; Roshani, G.H.; Nazemi, B.; Corniani, E.; Nazemi, E. Proposing a gamma radiation based intelligent system for simultaneous analyzing and detecting type and amount of petroleum by-products. *Neural Eng. Technol.* **2021**, *53*, 1277–1283. [CrossRef]
75. Wang, Y.; Cheng, H.; Hu, Q.; Liu, L.; Jia, L.; Gao, S.; Wang, Y. Pore structure heterogeneity of Wufeng-Longmaxi Shale, Sichuan Basin, China: Evidence from gas physisorption and multifractal geometries. *J. Petroleum Sci. Eng.* **2022**, *208*, 109313. [CrossRef]
76. Roshani, M.; Phan, G.; Roshani, G.H.; Hanus, R.; Nazemi, B.; Corniani, E.; Nazemi, E. Combination of X-ray tube and GMDH neural network as a nondestructive and potential technique for measuring characteristics of gas-oil-water three phase flows. *Measurement* **2021**, *168*, 108427. [CrossRef]
77. Xu, G.F. Numerical simulation and factor analysis of petrochemical pipe erosion-corrosion failure. In *IOP Conference Series: Materials Science and Engineering*; IOP Publishing: Bristol, UK, 2016; Volume 129, p. 012033.
78. Dolgii, I.E. Methods to enhance oil recovery in the process of complex field development of the Yarega oil and titanium deposit. *J. Min. Inst.* **2017**, *231*, 263–297.
79. Aleksandrov, A.N.; Kishchenko, M.A.; Van, T.N. Simulating the formation of wax deposits in wells using electric submersible pumps. In *Advances in Raw Material Industries for Sustainable Development Goals*; CRC Press: London, UK, 2021; pp. 283–295.
80. Nguyen, V.T.; Rogachev, M.K.; Aleksandrov, A.N. A new approach to improving efficiency of gas-lift wells in the conditions of the formation of organic wax deposits in the dragon field. *J. Pet. Explor. Prod. Technol.* **2020**, *10*, 3663–3672. [CrossRef]
81. Sandyga, M.S.; Struchkov, I.A.; Rogachev, M.K. Formation damage induced by wax deposition: Laboratory investigations and modeling. *J. Pet. Explor. Prod. Technol.* **2020**, *10*, 2541–2558. [CrossRef]
82. Sultanbekov, R.; Islamov, S.; Mardashov, D.; Beloglazov, I.; Hemmingsen, T. Research of the influence of marine residual fuel composition on sedimentation due to incompatibility. *J. Mar. Sci. Eng.* **2021**, *9*, 1067. [CrossRef]
83. Kashnikov, Y.A.; Ashikhmin, S.G.; Kukhtinskii, A.E.; Shustov, D.V. The relationship of fracture toughness coefficients and geophysical characteristics of rocks of hydrocarbon deposits. *J. Min. Inst.* **2020**, *1*, 241. [CrossRef]
84. Karpikov, A.V.; Aliev, R.I.; Babyr, N.V. An analysis of the effectiveness of hydraulic fracturing at YS1 of the Northern field. In *IOP Conference Series: Materials, Science and Engineering*; IOP Publishing: Bristol, UK, 2020; Volume 952, p. 012036.
85. Belonogov, E.V.; Korovin, A.Y.; Yakovlev, A.A. Increase of the injectivity coefficient by dynamic development of injection wells. *J. Min. Inst.* **2019**, *1*, 238.
86. Grigorev, M.B.; Tananykhin, D.S.; Poroshin, M.A. Sand management approach for a field with high viscosity oil. *J. Appl. Eng. Sci.* **2020**, *18*, 64–69. [CrossRef]
87. Shagiakhmetov, A.M.; Podoprigora, D.G.; Terleev, A.V. The study of the dependence of the rheological properties of gelforming compositions on the crack opening when modeling their flow on a rotational viscometer. *Period. Tche Quimica* **2020**, *17*, 933–939. [CrossRef]
88. Galkin, S.V.; Kochnev, A.A.; Zotikov, V.I. Predictive assessment of the effectiveness of radial drilling technology for the Bashkir production facilities of the Perm Territory fields. *J. Min. Inst.* **2019**, *1*, 238.
89. Molchanov, A.A.; Ageev, P.G. Implementation of new technologies is a reliable way of extracting residual reserves of hydrocarbon deposits. *J. Min. Inst.* **2017**, *1*, 227.
90. Prischepa, O.M.; Nefedov, Y.V.; Kochneva, O.E. Raw material base of hard-to-extract oil reserves of Russia (Matéria-prima base de reservas de oleo de dificil extração da Rússia). *Periodico TCHE Quimica* **2020**, *17*, 915–924.
91. Morenov, V.; Leusheva, E.; Martel, A. Investigation of the fractional composition effect of the carbonate weighting agents on the rheology of the clayless drilling mud. *Int. J. Eng.* **2018**, *31*, 1152–1158.
92. Nikitin, M.N.; Saychenko, L.A. The rheological properties of abnormally viscous oil. *Pet. Sci. Technol.* **2018**, *36*, 136–140. [CrossRef]
93. Ghanbari, B. Abundant exact solutions to a generalized nonlinear Schrödinger equation with local fractional derivative. *Math. Methods Appl. Sci.* **2021**, *44*, 8759–8774. [CrossRef]

94. Kharazmi, O.; Jahangard, S. A new family of lifetime distributions in terms of cumulative hazard rate function. *Commun. Fac. Sci. Univ. Ankara Series A1Math. Stat.* **2020**, *69*, 1–22. [CrossRef]
95. Ghanbari, B. Chaotic behaviors of the prevalence of an infectious disease in a prey and predator system using fractional derivatives. *Math. Methods Appl. Sci.* **2021**, *44*, 9998–10013. [CrossRef]
96. Kharazmi, O.; Saadatinik, A.; Jahangard, S. Odd hyperbolic cosine exponential-exponential (OHC-ee) distribution. *Annals Data Sci.* **2019**, *6*, 765–785. [CrossRef]
97. Ghanbari, B. On novel nondifferentiable exact solutions to local fractional Gardner's equation using an effective technique. *Math. Method Appl. Sci.* **2020**, *44*, 4673–4685. [CrossRef]
98. Rashidisabet, H.; Thomas, P.J.; Ajilore, O.; Zulueta, J.; Moore, R.C.; Leow, A. A systems biology approach to the Digital behaviorome. *Curr. Opin. Syst. Biol.* **2020**, *20*, 8–16. [CrossRef]
99. Nabti, A.; Ghanbari, B. Global Stability Analysis of a fractional SVEIR epidemic model. *Math. Methods Appl. Sci.* **2021**, *44*, 8577–8597. [CrossRef]
100. Rezaei, T.; Aslmarand, S.M.; Snyder, R.; Khajavi, B.; Alsing, P.M.; Fanto, M.; Ahn, D.; Miller, W.A. Experimental realization of Schumacher's information geometric Bell inequality. *Physics Lett. A* **2021**, *405*, 127444. [CrossRef]
101. Ghanbari, B. A fractional system of delay differential equation with nonsingular kernels in modeling hand-foot-mouth disease. *Adv. Diff. Eq.* **2020**, 2020. [CrossRef]
102. Ghanbari, B. On approximate solutions for a fractional prey-predator model involving the Atangana-Baleanu derivative. *Adv. Diff. Eq.* **2020**, *2020*, 1–20. [CrossRef]
103. Tian, J.; Liu, Y.; Zheng, W.; Yin, L. Smog prediction based on the deep belief - BP neural network model (DBN-BP). *Urban Climate* **2022**, *41*, 101078. [CrossRef]
104. Daryayehsalameh, B.; Nabavi, M.; Vaferi, B. Modeling of CO2 capture ability of [bmim][bf4] ionic liquid using connectionist smart paradigms. *Environ. Technol. Innov.* **2021**, *22*, 101484. [CrossRef]
105. Zhao, X.; Xia, H.; Pan, L.; Song, H.; Niu, W.; Wang, R.; Li, R.; Bian, X.; Guo, Y.; Qin, Y. Drought Monitoring over Yellow River Basin from 2003-2019 Using Reconstructed MODIS Land Surface Temperature in Google Earth Engine. *Remote Sens.* **2021**, *13*, 3748. [CrossRef]
106. Etimita, O.O.; Beka, F.T. Heavy mineral analysis of Eocene Sands and sandstones of Nanka Formation, Cenozoic Niger delta petroleum province. *Geol. Ecol. Landsc.* **2019**, *4*, 251–256. [CrossRef]
107. Singh, B.; Sihag, P.; Parsaie, A.; Angelaki, A. Comparative analysis of artificial intelligence techniques for the prediction of infiltration process. *Geol. Ecol. Landsc.* **2020**, *5*, 109–118. [CrossRef]
108. Taylor, J.G. *Neural Networks and Their Applications*; John Wiley & Sons Ltd.: Brighton, UK, 1996.
109. Gallant, A.R.; White, H. On Learning the Derivatives of an Unknown Mapping with Multilayer Feedforward Networks. *Neural Networks* **1992**, *5*, 129–138. [CrossRef]

applied sciences

MDPI

Article

Hydrodynamic Fingering Induced by Gel Film Formation in Miscible Fluid Systems: An Experimental and Mathematical Study

Muhammad Nasir [1], Ryuhei Yamaguchi [1], Yun She [1], Anindityo Patmonoaji [1,2,*], Mohammad Azis Mahardika [1,3], Weicen Wang [1], Zijing Li [1], Shintaro Matsushita [1] and Tetsuya Suekane [1,*]

1 Department of Mechanical Engineering, Tokyo Institute of Technology, 2-12-1-I6-33, Ookayama, Meguro-ku, Tokyo 152-8550, Japan; nasir.m.aa@m.titech.ac.jp (M.N.); mjuhvbn@outlook.jp (R.Y.); she.y.ab@m.titech.ac.jp (Y.S.); mahardika.m.aa@m.titech.ac.jp (M.A.M.); wang.w.ah@m.titech.ac.jp (W.W.); li.z.ax@m.titech.ac.jp (Z.L.); matsushita.s.ad@m.titech.ac.jp (S.M.)

2 Department of Chemical Engineering, Tokyo University of Agriculture and Technology, Naka-cho 2-24-16, Koganei, Tokyo 184-8588, Japan

3 Mechanical Engineering, Institut Teknologi Nasional Bandung, Bandung 40124, Indonesia

* Correspondence: patmonoaji.a.aa@m.titech.ac.jp (A.P.); suekane.t.aa@m.titech.ac.jp (T.S.); Tel.: +81-03-5734-3866 (A.P. & T.S.)

Citation: Nasir, M.; Yamaguchi, R.; She, Y.; Patmonoaji, A.; Mahardika, M.A.; Wang, W.; Li, Z.; Matsushita, S.; Suekane, T. Hydrodynamic Fingering Induced by Gel Film Formation in Miscible Fluid Systems: An Experimental and Mathematical Study. *Appl. Sci.* **2022**, *12*, 5043. https://doi.org/10.3390/app12105043

Academic Editor: Liangxing Li

Received: 22 April 2022
Accepted: 16 May 2022
Published: 17 May 2022

Abstract: Hydrodynamic fingering induced by gel formation shares common features with growing biofilms, bacterial colonies, and the instability of a confined chemical garden. Fluid displacement with gel formation is also essential in various engineering applications, including CO_2 leakage remediation from storage reservoirs and enhanced oil recovery. We conducted Hele-Shaw cell displacement experiments for a miscible fluid system using skim milk and aqueous citric acid solution. This study aimed to investigate the effects of gel film formation on the fingering instability of a miscible fluid system and develop a mathematical model of the sequential growth of gel film formation at the fingertip. We found that the gel film formation thickens with time, resulting in instability at the interface. A distinctive fingering pattern, resembling tentacles, appears where miscibility is suppressed, and the growth of the finger is localized at the fingertip. The finger width remains constant with increasing flow rate, whereas the number of fingers increases linearly before the fingers merge. The gap width significantly limits the finger width. Finally, a mathematical model of sequential film thickness growth for a bubble-like fingertip structure was developed. This model is based upon the interplay between the diffusion of citric acid through the existing gel film formation and elongation of the fingertip. The model provides an understanding of the fundamental mechanism of the growth of the bubble-like fingertip.

Keywords: gel film; finger instability; miscible; finger number; film thickness; finger width; mathematical model

1. Introduction

Phenomena associated with nonlinear dynamic instability are prevalent in nature [1–6]. Hydrodynamic fingering is a type of instability that occurs when one fluid is injected into another due to the change in fluid properties at bulk or interfacial regions, such as Saffman–Taylor instability, which is driven by viscosity difference [7–11]; Rayleigh–Taylor instability, which is driven by density difference [12–14]; and Marangoni instability, which is driven by interfacial tension difference [15–17]. Hydrodynamic fingering also occurs when a gel-like material forms at the interface of fluid pairs. Podgorski et al. [18] studied fingering instabilities that appear when two aqueous solutions of identical viscosities react together. A variety of fingering instabilities were observed owing to viscoelastic micellar formation at the interface between the two fluids. Some of the observed displacement patterns were similar to the instabilities of growing filamentary organisms. The growth of filamentary organisms was influenced significantly by a stretchable elastic membrane, resulting in a finger-like growth pattern [19]. Moreover, the growth of bacterial colonies through diffusion-limited processes in a nutrient concentration field results in the formation of random fractals, which have properties identical to the fingering pattern induced by gel formation [20–23].

The fingering instability induced by gel formation also shares mechanisms with precipitation patterns in confined chemical gardens, which are plant-like mineral structures originating from a precipitation reaction between a metallic salt and an aqueous alkaline solution. Precipitation is concentrated at the interface of reacted substances, forming a semipermeable membrane, out of which fingers grow, depending on the reactant concentrations. The spatial precipitate patterns depend on the injection direction, owing to the reaction–diffusion fronts that are asymmetrical [24–26].

Fluid displacement with gel formation is also essential in various engineering applications. Syed et al. [27] used a polyacrylamide gel with chromium as a crosslinker to study the use of polymer gel solutions for the remediation of potential CO_2 leakage from storage reservoirs. A similar study by Ito et al. [28] showed that injecting a silicate solution into a porous medium saturated with CO_2 results in the precipitation of amorphous silica that solidifies into a gel-like structure. This gel formation led to a 99% reduction in permeability and, thus, prevented CO_2 leakage. Furthermore, the gel blobs produced from a chemical reaction play a dominant role in improving oil recovery in the enhanced heavy oil recovery method based on alkali flooding. The gel blobs stick in the pores, blocking the flow path of the injected fluid. The blocked path causes the fluid to divert and push the trapped oil, thus, improving the oil recovery [29,30].

Hydrodynamic fingering is generally studied in Hele-Shaw cells, which are two parallel glass plates separated by a small gap that makes it possible to observe a two-dimensional flow pattern [31,32]. Moreover, the two-dimensional system is more accessible for numerical calculations. Flows in the Hele-Shaw cell can be modeled mathematically based upon Darcy's law [8]. Recently, a numerical simulation based on Computational Fluid Dynamics, micro-Particle Image Velocimetry, and Stereoscopic Particle Image Velocimetry have also been used to study the entrance phenomena, i.e., flow detachment and vortex generation in Hele-Shaw cells [33,34].

Although many experimental and theoretical studies have been conducted to study hydrodynamic fingering, quantitative studies on hydrodynamic fingering induced by gel formation, such as ones in which the effects of the sequential growth of the gel film thickness on fingering instability are analyzed, have not been performed and modeled mathematically. The gel film thickness significantly influences the instability of the displacement pattern. Moreover, other features of finger patterns, such as the finger number and width, were not examined. These features are expected to depend on parameters, such as the injection rate and gap width.

In this study, we aimed to (1) investigate the effects of gel film formation on the fingering instability of miscible fluids and (2) develop a mathematical model of the sequential growth of gel film formation at the tip of the finger. The experiments were conducted using a Hele-Shaw cell. An aqueous citric acid solution and skim milk were used as the fluid pair. The mixing of the two fluids resulted in gel formation, resembling the isoelectric coagulation of the yogurt fermentation process [35,36]. Moreover, the two fluids have similar viscosities to ensure that displacement instability does not occur unless there is gel formation.

The paper is organized as follows. Details of the experimental setup, fluid pair, and experimental procedures were given in Section 2. In Section 3, the typical structure of the fingering pattern induced by gel formation was discussed, and the diffusion coefficient was estimated. The changes in displacement pattern with parameters, such as citric acid concentration, injection rate, and the Hele-Shaw cell gap, were captured in phase diagrams. In addition, the displacement pattern upon reverse injection was assessed. The sequential growth of gel film formation at the fingertip was modeled mathematically. Finally, the findings were summarized in Section 4.

2. Methods

2.1. Experimental Setup

Figure 1 shows a schematic diagram of the experimental setup. The experiments were conducted using a Hele-Shaw cell, a device that has a small gap between two parallel glass plates that makes it possible to observe a two-dimensional flow pattern [37]. The area of the two square glass plates was 300 × 300 mm. A square copper block was placed as a spacer at each edge of the parallel glass plates. These spacers made it possible to adjust the gap width b between the two plates. In this work, the gap width b was varied from 0.4 to 0.8 mm to analyze its effect on the displacement patterns. The lower glass plate was equipped with a hole in the center for liquid injection. A camera Canon EOS 60D (Canon Inc., Tokyo, Japan) was installed at the top of the Hele-Shaw cell, and a video was taken when the displacing fluid was injected. A syringe pump KDS100 (KD Scientific Inc., Holliston, MA, USA) was used to inject fluid. In addition, a torch ESL-W2001AC (Asahi Electric Co., Ltd., Osaka, Japan) was located at the bottom of the experimental device to illuminate the Hele-Shaw cell so that shadows were not reflected in the recorded video.

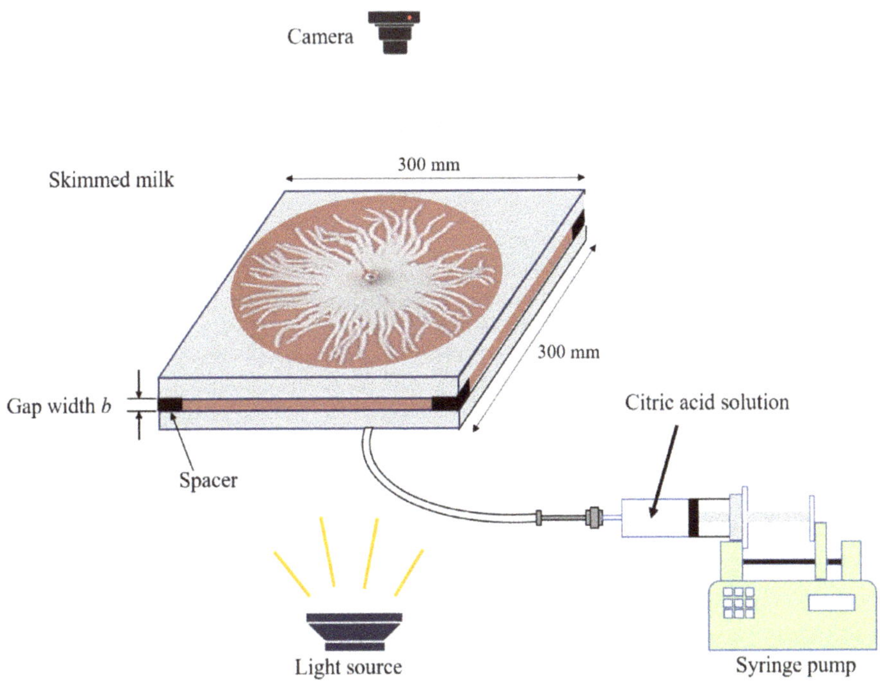

Figure 1. Schematic diagram of the experimental setup.

2.2. Fluid Pair

The fluids used in this experiment were skim milk and aqueous citric acid solution. Skim milk is a nonfat milk that contains a relatively high amount of protein, of which approximately 80% is casein. Casein within skim milk is stable because the milk has a pH value of approximately 6.60. However, when acid is added to the milk, casein becomes unstable as the pH drops, approaching the isoelectric point (the least soluble pH value) of casein, which is at pH 4.6, resulting in isoelectric precipitation [38]. The aggregation of casein micelles resulting from isoelectric precipitation subsequently leads to viscoelastic gel formation, a typical gel formation found in the yogurt fermentation process [35,36].

The skim milk and the aqueous citric acid solution have an identical viscosity η of 2.07 mPa·s and 1.04–1.59 mPa·s, respectively. A low-viscosity ratio eliminates displacement instability caused by the viscosity difference [39]. The density ρ of the skim milk was 1.01 kg/m^3 measured at a temperature of 22.80 °C. Skim milk powder was dissolved in water at a ratio of 10 g skim milk powder to 100 g of water. The protein content of this skim milk powder is 3.40 g for 10 g. Therefore, using weight percent (wt%) concentration, which is based on solute to solution ratio, the protein concentration of the solution used in this experiment was 3.1 wt%. The concentration was kept constant throughout the experiment. The food coloring mixtures were added to skim milk at a ratio of 0.10 g to 100 g of water to differentiate the skim milk and the aqueous citric acid solution. The skim milk was also doped with a methyl red pH indicator in several experiments to clearly observe the region where the gel emerged. Methyl red is a pH indicator that is yellow at a pH of more than 6 (skim milk pH value) and red at a pH of less than 4.6 (at which casein micelles precipitate) [40].

There were five concentrations of citric acid used in this experiment: 0.0 (pure water), 0.50, 1.5, 5.0, and 15 wt%. Table 1 shows the measured values for the density ρ, dynamic viscosity η, concentration c, and temperature T of citric acid.

Table 1. Measured values of density ρ, viscosity η, and temperature T at each concentration c of the citric acid used in the experiment.

c [wt%]	ρ [kg/m^3]	η [mPa·s]	T [°C]
0.0	1.00	1.17	18.8
0.5	0.99	1.04	21.8
1.5	0.99	1.12	21.1
5.0	1.00	1.18	20.7
15	1.05	1.59	22.8

2.3. Experimental Procedures

Two different types of experiments were conducted. In the first experiment, the displaced fluid was skim milk, whereas the displacing fluid was the aqueous citric acid solution (direct injection). The second experiment was a reverse injection experiment, in which the displaced and displacing fluids were exchanged accordingly. In both experiments, the Hele-Shaw cell was first saturated with the displaced fluid. After complete saturation, the displacing fluid was injected through a hole in the center of the lower glass plate of the Hele-Shaw cell. Displacement patterns were recorded using a digital camera. The experiment was stopped when the injected fluid volume reached 2 mL. The same procedure was repeated for experiments with different citric acid concentrations and injection rates Q (15–500 mL/h).

3. Results and Discussion

3.1. Typical Fingering Structure Induced by Gel Formation

Figure 2a shows the fingering pattern that emerges when aqueous citric acid solution displaces skim milk. The remarkable feature of this finger pattern is the presence of gel formation (red) that emerges at the interface between the skim milk and the aqueous citric acid solution. The gel formation acts as a continuous sidewall, preventing the finger from branching; thus, only the tip of the elongated finger continues to grow. This typical characteristic of the fingering pattern differentiates the fingering resulting from gel formation and the fingering caused by the viscosity difference [9].

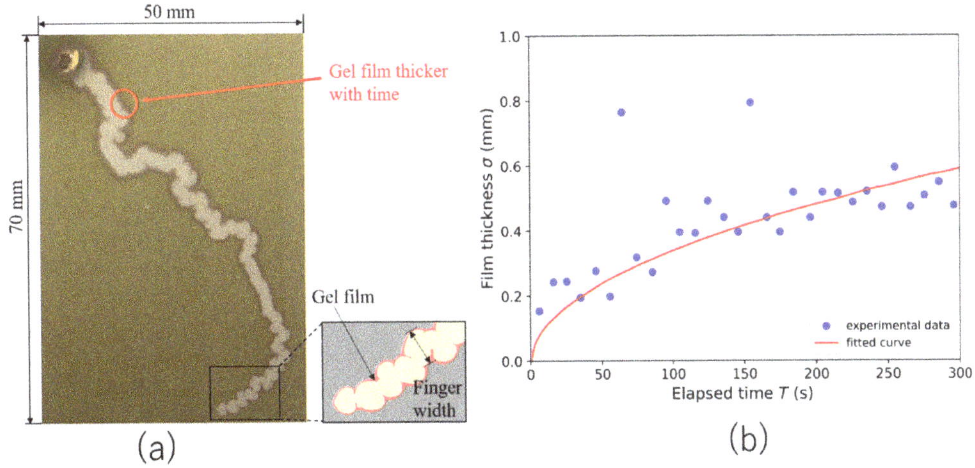

Figure 2. (**a**) A typical finger structure consists of bubble-like patterns with gel formation (red) generated at a 15 wt% concentration of the citric acid, an injection rate Q of 1.0 mL/h, and a gap width b of 0.4 mm. The skim milk was doped with a methyl red pH indicator. (**b**) The sequential thickening of the gel film at the interface of the skim milk and aqueous citric acid solution. The blue dot represents the experimental result, whereas the red line is the fitting curve based on Equation (1).

The gel film formation thickens with time at the interface between the aqueous citric acid solution and skim milk. Figure 2b shows the sequential thickening of the gel film. The gel film thickness increases logarithmically with time. The gel film thickens faster in the early stage and gradually becomes constant. The thickening of the gel film is caused by the diffusion of the citric acid through the gel film formation, resulting in further mixing of skim milk and citric acid. The relation between the film thickness σ growth over time t and the diffusion coefficient constant D is expressed as follows [41]:

$$\sigma = \sqrt{4\pi D t} \tag{1}$$

We used the least-squares method to estimate the diffusion coefficient D that best fit the observed data, excluding two outlier data points ($\sigma > 0.76$). The estimated D is 9.3×10^{-11} m^2/s, with the coefficient of determination (R-squared), accounting for 72.5%. The estimated D is lower than the diffusion coefficient in a normal solution due to the bending of the diffusion path owing to gel formation. Moreover, unlike the diffusion phenomenon in uniform systems, the diffusion coefficient through the gel formation layer is not constant and depends on the solute concentration in the gel [42].

3.2. Effects of Injection Rate and Concentration of the Citric Acid on Displacement Pattern

We compared the fingering patterns that appeared under each condition when the injection rate and concentration of the citric acid were varied. Figure 3 shows the fingering patterns when citric acid is injected into skim milk under different injection rates and citric acid concentrations. Instability does not occur when pure water is used because no gel forms. However, the displacement becomes unstable when the citric acid concentration increases; thus, fingers emerge. The fingering pattern occurs because of casein precipitation at the interface, as the pH value decreases with increasing citric acid concentration. However, the displacement pattern becomes more stable with an increase in the injection rate.

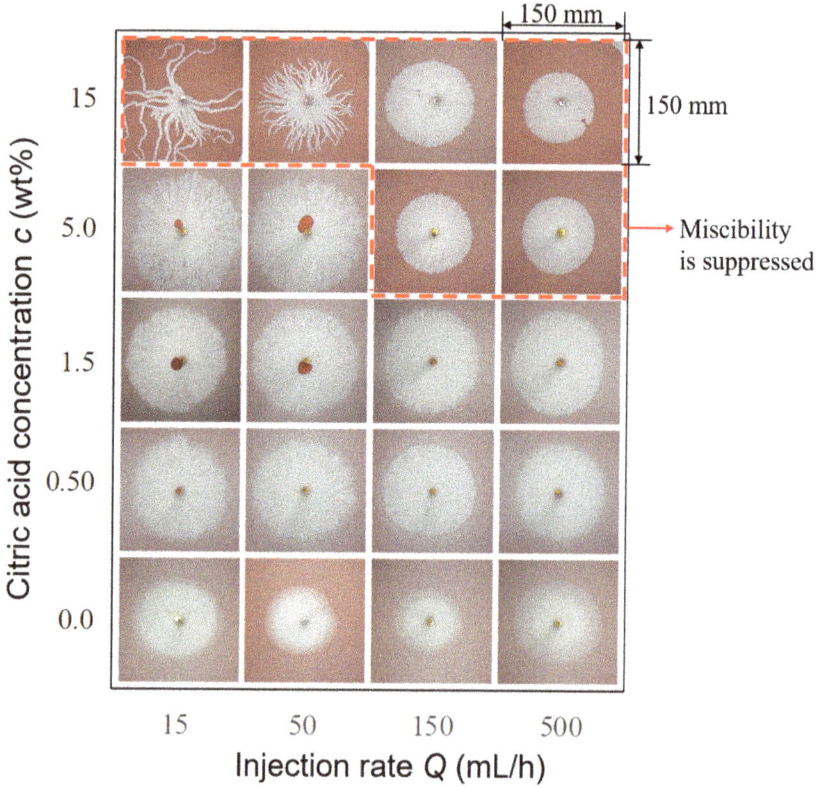

Figure 3. Displacement patterns when citric acid (white) is injected into skim milk (pink) upon injection rates of 15, 50, 150, and 500 mL/h and citric acid concentrations of 0.0 (pure water), 0.50, 1.5, 5.0, and 15 wt%. Images are taken when the injection volume of the aqueous citric acid solution is 2 mL and the gap width b is 0.4 mm. The red dash line represents the conditions at which the miscibility of citric acid and skim milk is suppressed.

The mixing between the skim milk and the aqueous citric acid solution occurs when the concentrations of the citric acid are 0.0, 0.50, 1.5 (all injection rates), and 5.0 wt% (injection rates of 15 and 50 mL/h). The mixing part is represented by the color gradient between white and pink at the interface of the fluid pair. At a low concentration of citric acid, the miscibility is not suppressed because of insufficient gel formation at the interface of the fluid pair; thus, mixing occurs. Moreover, the small amount of gel formation enables branching of the finger to occur at the sidewall of the fingers, mimicking the displacement pattern of viscous fingering.

However, at a high citric acid concentration of 15 wt% (all injection rates) and 5 wt% (injection rates of 150 and 500 mL/h), the mixing between the skim milk and the aqueous citric acid solution is suppressed. A high concentration of citric acid results in high casein precipitation, which turns into a large amount of gel formation and becomes a continuous wall at the interface between skim milk and aqueous citric acid solution. This gel formation hinders the miscibility of the two fluids.

Moreover, unlike the finger pattern that emerges when the aqueous citric acid solution is mixed with skim milk, remarkable fingering patterns, identical to tentacle-like structures, appear in the state where miscibility is suppressed (at a citric acid concentration of 15 wt% and injection rates of 15 and 50 mL/h). The number of branches on the finger is extremely small. The gel formation prevents the finger from branching, as previously mentioned;

hence, the growth of the finger is localized at the tip. When the finger reaches a certain length, the growth stops, and a new finger is generated from the base of the finger.

Figure 4 shows how the number of finger changes upon different injection rates Q, at a citric acid concentration of 15 wt%. Increasing the injection rate up to an injection rate of 200 mL/h leads to a linear increase in finger numbers. However, although the number of fingers increases, the finger width is constant. As a result, more fingers emerge to compensate for a larger injected volume of aqueous citric acid solution. When the injection rate is higher than 200 mL/h, the distance between the fingers becomes closer as the number of fingers increases. At this point, the number of fingers cannot increase anymore unless the finger width becomes small or the displacement front has traveled far enough from the injection point. As a result, the finger is likely to merge, denoted as the transition region in Figure 4. Further increasing the injection rate, therefore, decreases the total number of fingers.

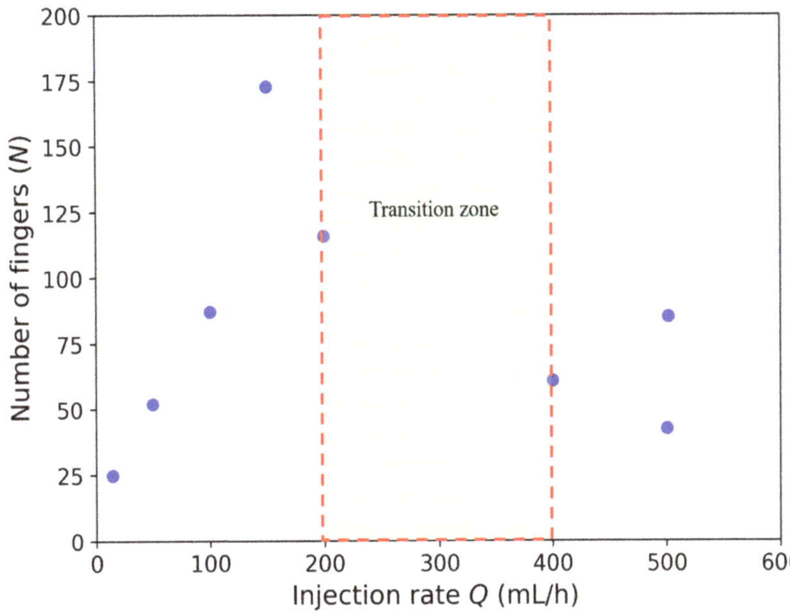

Figure 4. Relationship of injection rate Q with respect to the number of fingers N at a citric acid concentration c = 15 wt% (when the volume of the injected aqueous citric acid solution is 2 mL and the gap width b is 0.4 mm). The transition zone represents the region at which the fingers begin to merge.

3.3. Gap Width Effects on the Finger Width

We expect that the finger width is significantly restricted by the gap width b. Therefore, we varied the gap width b from 0.4 to 0.8 mm and analyzed the results. Figure 5a shows a diagram of fingering patterns under four conditions with the change in gap width b and injection rates. In both gap width b conditions, the number of fingers increases when the injection flow rate increases, which is the same result as that discussed in the previous section. The typical fingering pattern is similar even when the gap width changes. However, finger width is reduced when the gap width b is smaller, whereas increasing the gap width b leads to a larger finger width. A larger finger width is accompanied by a decrease in the number of fingers. Figure 5a shows that approximately 25 fingers appear at a gap width b of 0.4 mm and an injection rate of 15 mL/h. However, at the same injection rate, the number of fingers decreases to only five when the gap width b is 0.8 mm. The decrease in the number of fingers compensates for the increased finger width at the same volume of the injected aqueous citric acid solution.

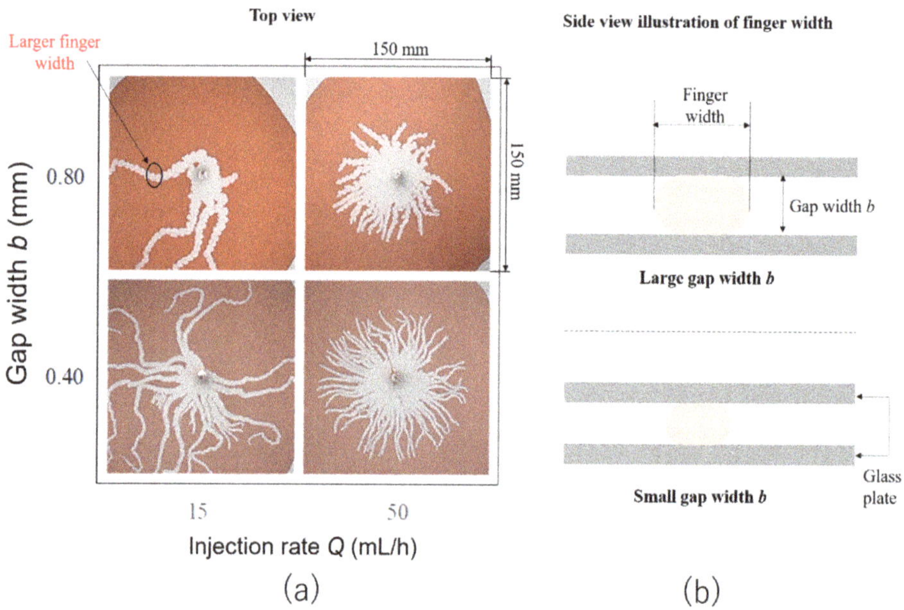

Figure 5. (**a**) Top view of displacement patterns when the gap widths are 0.4 and 0.8 mm and injection rates were varied from 15 to 50 mL/h. The volume of the injected aqueous citric acid solution is 2 mL. (**b**) Side view illustration of how the finger width changes with gap width b. The sphere-like shape represents the aqueous citric acid solution.

The proportional relation between the gap width b and finger width is subject to the tendency of the aqueous citric acid solution to have a sphere-like shape because of surface tension. When the gap width b is wider, the sphere-like shape of the aqueous citric acid solution also becomes larger and, thus, the finger width, as illustrated in Figure 5b. When the gap width b is smaller, however, the finger width also decreases. This result confirms that the finger width is significantly limited by the gap width b.

3.4. *Reverse Injection Displacement Pattern (Skim Milk Injected into Citric Acid)*

We also conducted experiments in which the displacing and displaced fluids were exchanged. The aqueous citric acid solution was first saturated in the Hele-Shaw cell as a displaced fluid, and skim milk was then injected as a displacing fluid. Figure 6 shows displacement patterns when the skim milk is injected into citric acid under different injection rates and citric acid concentrations. Stable displacement appears at a higher injection rate, whereas a higher citric acid concentration leads to displacement instability. The observed trend is similar to that when the fluid pair is not reversed. However, the displacement patterns are different. The growing front of the displacement pattern has a flower-like appearance of wide fingers. Such patterns might occur because of the asymmetry in the diffusion coefficients of the two mixing species. The precipitated casein likely has higher mobility in water (skim milk), reducing overall gel formation at the interface [18]. Resistance to finger widening was also reduced because there was a low amount of gel formation at the interface of the skim milk and aqueous citric acid solution.

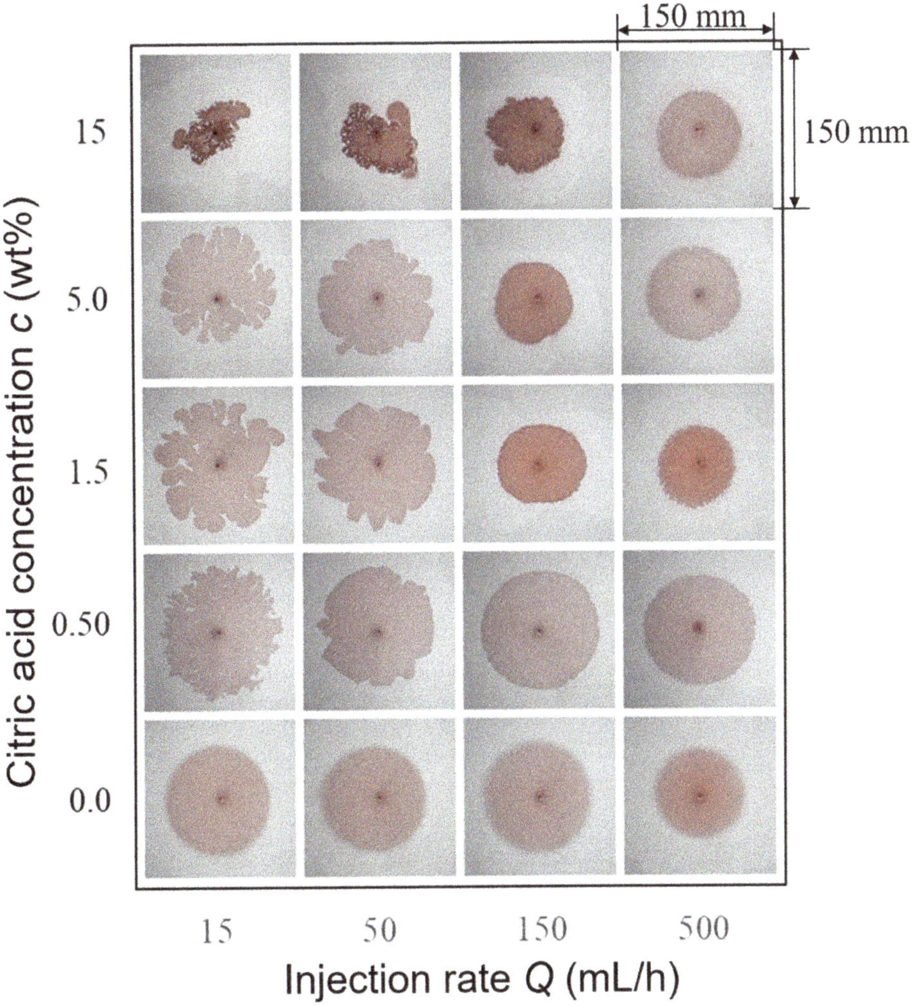

Figure 6. Displacement patterns when skim milk (pink) is injected into citric acid (white) upon injection rates of 15, 50, 150, and 500 mL/h and citric acid concentrations of 0.0 (pure water), 0.50, 1.5, 5.0, and 15 wt%. Images were taken when the injection volume was 2 mL, and the gap width b was 0.4 mm.

Gel formation occupies the whole skim milk, at a citric acid concentration of 15 wt%. The area displaced by skim milk appears smaller (colored area), even though the injection volume is the same. This result suggests that the water from the milk leaks into citric acid and appears transparent, whereas gel formation accumulates close to the injection source.

To further confirm the mixing of water from milk with citric acid, additional experiments were performed by adding methyl red to the skim milk side. The displacement pattern when the skim milk was doped with the methyl red pH indicator is shown in Figure 7. The pink part is the mixing region where the water was contained in skim milk combined with the aqueous citric acid solution. This pink region is not observable in Figure 6; however, with the help of the methyl red pH indicator, this mixing region can be observed in Figure 7. The dark red indicates gel formation resulting from casein micelle precipitation. The dark red part spreads over the entire region of skim milk, which confirms that the precipitation of casein micelles is asymmetric and accumulated throughout the skim milk. Even at a high citric acid concentration, with a large amount of gel formation, mixing between skim milk and citric acid occurs because the gel does not solidify at the interface. Therefore, it can be inferred that the asymmetry of gel formation greatly affected the different displacement patterns between the reversed injection and direct injection experiments.

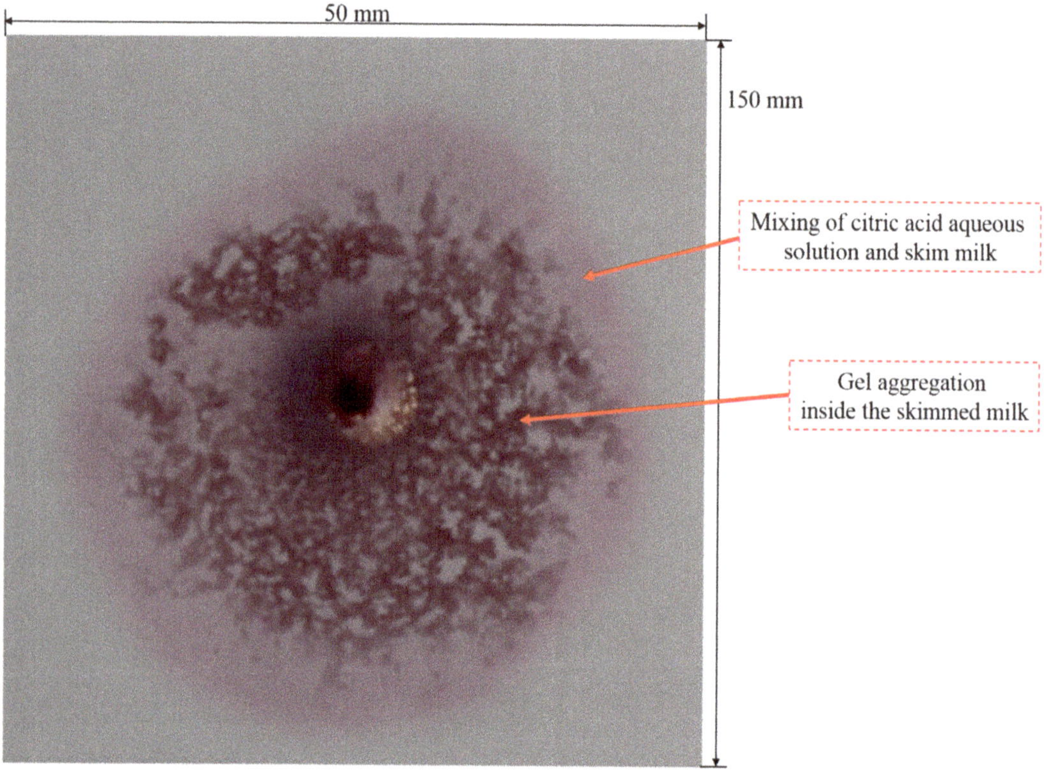

Figure 7. Displacement pattern when the skim milk doped with methyl red is injected into the aqueous citric acid solution. The dark red part is the gel resulting from precipitated casein micelles, whereas the pink part is the water separated from the gel. The image was captured at the initial injection stage when the skim milk was injected into the aqueous citric acid solution, and the skim milk injection volume was 0.4 mL. The citric acid concentration c was 15 wt% at an injection rate Q of 15 mL/h and a gap width b of 0.4 mm.

3.5. Mathematical Model of Sequential Growth of Gel Film Thickness at the Fingertip

Figure 8a shows the sequential growth of bubble-like fingers that emerges at a high citric acid concentration. The growth of this finger comprises a series of expansions, stagnations, and explosions that repeat over time. The pressure of the neighboring fingers does not influence the emerging bubble-like fingers. As shown in Figure 2a, even a single finger may possess this feature.

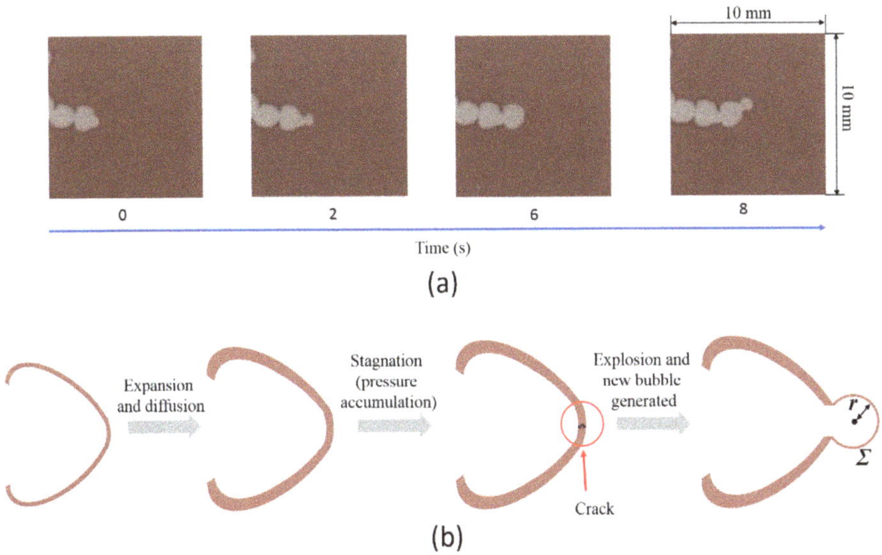

Figure 8. (**a**) Sequential growth of bubble-like finger structure that appears at a citric acid concentration c of 15 wt%, injection speed Q of 15 mL/h, and gap width b of 0.4 mm. (**b**) Illustration of the growth of fingertip comprising expansion, stagnation, and explosion.

The reason why this structure appeared is because of the change in the gel film thickness over time. The film thickness decreases when the fingertip is elongated. However, the diffusion of citric acid through the existing gel formation increases the gel film thickness. Figure 8b illustrates the growth of the fingertip. Initially, the fingertip expands, resulting in a reduction in gel film thickness. However, at the same time, diffusion overcomes the elongation effect, resulting in a thicker gel film over time, while the fingertip continues to expand. At a certain point, the expansion stops because the gel film formation is sufficiently thick to act against the injection pressure. When finger growth stops, the pressure continues to increase. The increased pressure causes the gel film to burst. After the rupture, the finger grows again, and the pressure returns to its original state, marking the cycle repetition.

An attempt was made to mathematically model the change in the film thickness at the tip of the finger when this phenomenon occurred. First, the increase in the film thickness per unit time resulting from diffusion was derived. The time change in the film thickness owing to diffusion is expressed by Equation (1), based on Fick's law. Differentiating Equation (1) for time, the increase in gel film thickness per unit time can be expressed as:

$$\frac{d\sigma}{dt} = \sqrt{\frac{\pi D}{t}} \tag{2}$$

Next, the decrease in gel film thickness per unit time caused by extension was derived. When the number of fingers is one, the shape of the fingertip of a newly emerged bubble-like tip is modeled as a circle, the radius is r, the injection velocity is q, and the cell gap width is b. Then, the following equation holds from the preservation of the volume increase in the injected fluid:

$$2\pi brdr - Qdt \tag{3}$$

Integrating both sides of the above equation, one obtains

$$\pi br^2 = Qt + C \tag{4}$$

Furthermore, if $t = 0$ and $r = r_0$, the above equation can be expressed as

$$\pi b r^2 = Qt + \pi b r_0^2 \tag{5}$$

Equation (3) can be rearranged as

$$\frac{dr}{dt} = \frac{Q}{2\pi b r} \tag{6}$$

If the thickness of the gel film is Σ, the following equation holds, according to the volume conservation law of the gel film:

$$2\pi b r \Sigma = \text{const} \tag{7}$$

Differentiating Equation (7) with time t, the following equation is obtained:

$$\frac{dr}{dt}\Sigma + r\frac{d\Sigma}{dt} = 0 \tag{8}$$

Based on Equations (5) and (6), Equation (8) can be expressed as

$$\frac{d\Sigma}{dt} = -\frac{1}{r}\frac{dr}{dt}\Sigma = -\frac{q}{2\pi b r^2}\Sigma = -\frac{1}{2\left(t + \frac{\pi b r_0^2}{Q}\right)}\Sigma \tag{9}$$

To obtain the overall change in film thickness with time, Equation (2) was combined with Equation (9), which represents the change in film thickness resulting from diffusion and fingertip expansion.

$$\frac{d\Sigma}{dt} = \sqrt{\frac{\pi D}{t}} - \frac{\Sigma}{2\left(t + \frac{\pi b r_0^2}{Q}\right)} \tag{10}$$

The above equation is transformed as follows for simplicity:

$$\frac{d\Sigma}{dt} = \sqrt{\frac{\pi D}{t + \frac{\pi b r_0^2}{q}}} - \frac{\Sigma}{2\left(t + \frac{\pi b r_0^2}{Q}\right)} \tag{11}$$

Furthermore, by solving Equation (11), the following equation is obtained:

$$\Sigma = \frac{\sqrt{\pi D}t + C}{\sqrt{t + \frac{\pi b r_0^2}{Q}}} \tag{12}$$

This equation makes it possible to model the growth of the gel film thickness at the tip of the finger from the initial time when the bubble-like shape begins to expand until the bubble bursts. Assuming that r_0 is 0.4 mm, which is the same as the gap width, and $C = 3.404 \times 10^{-5}$ m$\sqrt{\text{s}}$, the value such that the film thickness at $t = 0$ is 0.04 mm, a graph showing the sequential growth of film thickness at the tip of the finger can be obtained (Figure 9). At the initial state, $t = 0$, the film thickness decreases with the extension of the fingertip, whereas the effect of diffusion is insignificant. The film thickness reaches a minimum value at a certain point. After reaching the minimum point, the film thickness recovers and increases with time, owing to a significant diffusion effect. As the film thickness increases, the pressure at the fingertip is also expected to increase because finger extension is less likely to occur as the film thickness increases. As a result, the gel film cannot withstand the pressure, and a crack occurs at the tip of the gel film formation, resulting in a bubble-like finger burst. The tip of the finger repeatedly bursts at intervals of approximately 9 s, as expected, based on the experimental results shown in Figure 8a.

When a bubble bursts, a new bubble-like tip emerges. This typical cycle repeats over time as more bubble-like fingertips are continuously generated.

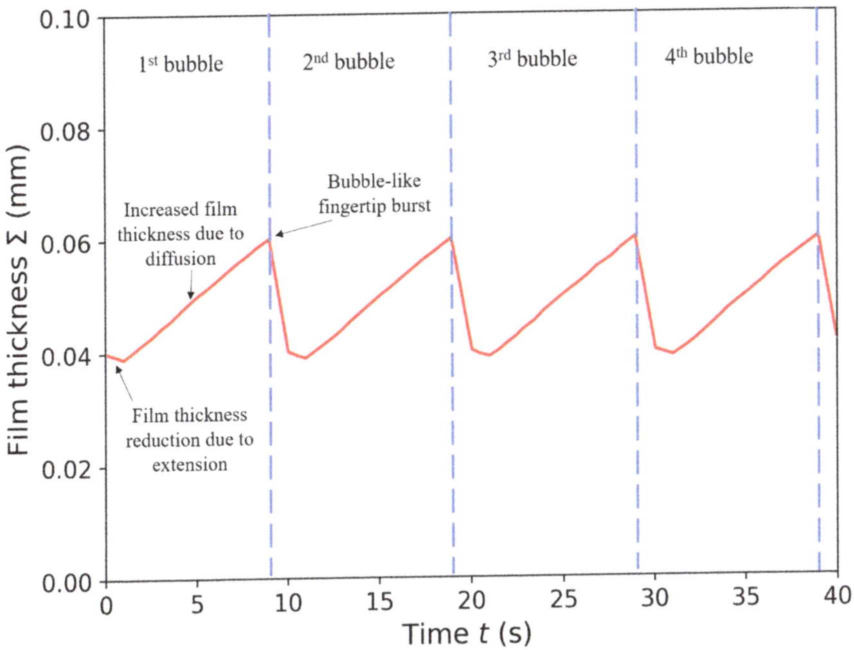

Figure 9. Sequential change in the interfacial gel film thickness based on Equation (12) considering periodic rupture at the fingertip. The parameters are set as follows: $D = 9.3 \times 10^{-11}$ m^2/s, $b = 0.4$ mm, $Q = 1.0$ mL/h, $r_0 = 0.4$ mm, and $C = 3.404 \times 10^{-5}$ m$\sqrt{s}$. The film thickness changes from decreasing to monotonically increasing as a one-cycle bubble formation. Periodic rupture occurs approximately every 9 s.

4. Conclusions

In this study, we conducted Hele-Shaw cell displacement experiments using skim milk and aqueous citric acid solution to investigate the effects of gel film formation on the fingering instability of miscible fluids and develop a mathematical model for the sequential growth of gel film formation at the fingertip. We found that mixing skim milk and an aqueous citric acid solution induces gel film formation, resulting in interface instability. A distinctive feature of the fingering pattern, such as tentacles, appears when miscibility is suppressed, and the growth of the finger is localized at the fingertip. The gel film formation thickens with time owing to the diffusion of citric acid through the existing gel film formation.

The finger width remains constant with increasing flow rate, whereas the number of fingers increases linearly. However, at a certain point, where the injection rate is too high to be compensated by the increase in the number of fingers, the fingers merge and the number of fingers decreases. The finger patterns that appear at each injection rate are similar, regardless of the gap width. However, the finger width increases with the gap width. In addition, the growing front of the displacement pattern has a flower-like appearance of wide fingers when the displaced and displacing fluids are exchanged.

The growth of the bubble-like fingertip comprises a series of expansions, stagnations, and explosions that repeats over time. A mathematical model of sequential film thickness growth for a bubble-like fingertip structure was developed. The model is based upon the interplay between the diffusion of citric acid through the existing gel film formation and elongation of the fingertip, which provides an understanding of the fundamental mechanism for the growth of the bubble-like fingertip, owing to gel film formation. This model is potentially relevant to various processes, including the growth of bacterial biofilms [19] and fingering instability in confined chemical gardens [24,25]. The results of this investigation can also be a reference for engineering applications, such as remediation of potential CO_2 leakage from storage reservoirs [27,28] and enhanced heavy oil recovery methods based on alkali flooding [29,30]. We believe that future research must be directed toward further analysis based on pore-scale studies using a porous medium structure that resembles the structure of geological formation. Pore-scale studies offer a detailed observation for the underlying mechanisms, which are macroscopically unobserved [43].

Author Contributions: Conceptualization, T.S. and R.Y.; methodology, M.N., R.Y. and Y.S.; validation, A.P., M.A.M. and Z.L.; formal analysis, M.N., R.Y. and W.W; investigation, R.Y., Y.S. and W.W.; writing—original draft preparation, M.N. and R.Y.; writing—review and editing, M.N., A.P. and M.A.M.; visualization, M.N. and Z.L.; supervision, S.M., A.P. and T.S.; project administration, S.M.; funding acquisition, A.P. and T.S. All authors have read and agreed to the published version of the manuscript.

Funding: This study was supported by the Japan Society for the Promotion of Science KAKENHI (grant numbers 17H00790 and 20J14975), InfoSyEnergy Program Tokyo Institute of Technology, and Indonesia Endowment Fund for Education (LPDP).

Institutional Review Board Statement: Not applicable.

Informed Consent Statement: Not applicable.

Data Availability Statement: The data presented in this study are available on request from the corresponding author.

Conflicts of Interest: The authors declare no conflict of interest.

References

1. Ghatak, A.; Chaudhury, M.K.; Shenoy, V.; Sharma, A. Meniscus Instability in a Thin Elastic Film. *Phys. Rev. Lett.* **2000**, *85*, 4329–4332. [CrossRef]
2. Sandnes, B.; Flekkoy, E.G.; Knudsen, H.A.; Maloy, K.J.; See, H. Patterns and flow in frictional fluid dynamics. *Nat. Commun.* **2011**, *2*, 288. [CrossRef] [PubMed]
3. Chaudhury, M.K.; Chakrabarti, A.; Ghatak, A. Adhesion-induced instabilities and pattern formation in thin films of elastomers and gels. *Eur. Phys. J. E* **2015**, *38*, 82. [CrossRef] [PubMed]
4. Sakai, S.; Nakanishi, Y.; Hyodo, A.; Wang, L.; Suekane, T. Three-dimensional fingering structure associated with gravitationally unstable mixing of miscible fluids in porous media. *Heat Transf. Res.* **2018**, *49*, 1023–1039. [CrossRef]
5. Nagatsu, Y.; Ishii, Y.; Tada, Y.; De Wit, A. Hydrodynamic Fingering Instability Induced by a Precipitation Reaction. *Phys. Rev. Lett.* **2014**, *113*, 024502. [CrossRef]
6. Patmonoaji, A.; Hu, Y.; Nasir, M.; Zhang, C.; Suekane, T. Effects of Dissolution Fingering on Mass Transfer Rate in Three-Dimensional Porous Media. *Water Resour. Res.* **2021**, *57*, e2020WR029353. [CrossRef]
7. Saffman, P.G. Viscous fingering in Hele-Shaw cells. *J. Fluid Mech.* **1986**, *173*, 73–94. [CrossRef]
8. Stewart, S.; Marin, D.; Tullier, M.; Pojman, J.; Meiburg, E.; Bunton, P. Stabilization of miscible viscous fingering by a step growth polymerization reaction. *Exp. Fluids* **2018**, *59*, 114. [CrossRef]
9. Sin, S.; Suekane, T.; Nagatsu, Y.; Patmonoaji, A. Three-dimensional visualization of viscous fingering for non-Newtonian fluids with chemical reactions that change viscosity. *Phys. Rev. Fluids* **2019**, *4*, 054502. [CrossRef]
10. Nagatsu, Y.; Kondo, Y.; Kato, Y.; Tada, Y. Effects of moderate Damköhler number on miscible viscous fingering involving viscosity decrease due to a chemical reaction. *J. Fluid Mech.* **2009**, *625*, 97–124. [CrossRef]
11. Sorbie, K.S.; Al Ghafri, A.Y.; Skauge, A.; Mackay, E.J. On the Modelling of Immiscible Viscous Fingering in Two-Phase Flow in Porous Media. *Transp. Porous Media* **2020**, *135*, 331–359. [CrossRef]
12. Hewitt, D.R.; Neufeld, J.A.; Lister, J.R. Convective shutdown in a porous medium at high Rayleigh number. *J. Fluid Mech.* **2013**, *719*, 551–586. [CrossRef]
13. Forbes, L.K.; Browne, C.A.; Walters, S.J. The Rayleigh–Taylor instability in a porous medium. *SN Appl. Sci.* **2021**, *3*, 188. [CrossRef]
14. Livescu, D.; Wei, T.; Brady, P. Rayleigh–Taylor instability with gravity reversal. *Phys. D Nonlinear Phenom.* **2021**, *417*, 132832. [CrossRef]
15. Wang, T.-S.; Shi, W.-Y. Marangoni instability induced by evaporation in well-defined non-spherical sessile droplet. *Int. J. Heat Mass Transf.* **2019**, *141*, 168–179. [CrossRef]

16. Li, Y.; Meijer, J.G.; Lohse, D. Marangoni instabilities of drops of different viscosities in stratified liquids. *J. Fluid Mech.* **2021**, *932*. [CrossRef]
17. Li, Y.; Diddens, C.; Prosperetti, A.; Lohse, D. Marangoni Instability of a Drop in a Stably Stratified Liquid. *Phys. Rev. Lett.* **2021**, *126*, 124502. [CrossRef]
18. Podgorski, T.; Sostarecz, M.C.; Zorman, S.; Belmonte, A. Fingering instabilities of a reactive micellar interface. *Phys. Rev. E-Stat. Nonlinear Soft Matter Phys.* **2007**, *76*, 016202. [CrossRef]
19. Goriely, A.; Tabor, M. Self-Similar Tip Growth in Filamentary Organisms. *Phys. Rev. Lett.* **2003**, *90*, 108101. [CrossRef]
20. Tronnolone, H.; Tam, A.; Szenczi, Z.; Green, J.E.F.; Balasuriya, S.; Tek, E.L.; Gardner, J.M.; Sundstrom, J.F.; Jiranek, V.; Oliver, S.; et al. Diffusion-Limited Growth of Microbial Colonies. *Sci. Rep.* **2018**, *8*, 5992. [CrossRef]
21. Matsushita, M.; Fujikawa, H. Diffusion-limited growth in bacterial colony formation. *Phys. A Stat. Mech. Its Appl.* **1990**, *168*, 498–506. [CrossRef]
22. Schwarcz, D.; Levine, H.; Ben-Jacob, E.; Ariel, G. Uniform modeling of bacterial colony patterns with varying nutrient and substrate. *Phys. D Nonlinear Phenom.* **2016**, *318–319*, 91–99. [CrossRef]
23. Mimura, M.; Sakaguchi, H.; Matsushita, M. Reaction–diffusion modelling of bacterial colony patterns. *Phys. A Stat. Mech. Its Appl.* **2000**, *282*, 283–303. [CrossRef]
24. Haudin, F.; Cartwright, J.H.E.; Brau, F.; De Wit, A.; Takayama, K.; Morisaki, Y.; Kuno, S.; Nagamoto, Y.; Harada, K.; Furukawa, N.; et al. Spiral precipitation patterns in confined chemical gardens. *Proc. Natl. Acad. Sci. USA* **2014**, *111*, 17363–17367. [CrossRef]
25. Haudin, F.; Cartwright, J.H.E.; De Wit, A. Direct and Reverse Chemical Garden Patterns Grown upon Injection in Confined Geometries. *J. Phys. Chem. C* **2015**, *119*, 15067–15076. [CrossRef]
26. Haudin, F.; Brasiliense, V.; Cartwright, J.H.E.; Brau, F.; De Wit, A. Genericity of confined chemical garden patterns with regard to changes in the reactants. *Phys. Chem. Chem. Phys.* **2015**, *17*, 12804–12811. [CrossRef]
27. Syed, A.; Pantin, B.; Durucan, S.; Korre, A.; Shi, J.Q. The use of polymer-gel solutions for remediation of potential CO_2 leakage from storage reservoirs. *Energy Procedia* **2014**, *63*, 4638–4645. [CrossRef]
28. Ito, T.; Xu, T.; Tanaka, H.; Taniuchi, Y.; Okamoto, A. Possibility to remedy CO_2 leakage from geological reservoir using CO_2 reactive grout. *Int. J. Greenh. Gas Control* **2014**, *20*, 310–323. [CrossRef]
29. Mahardika, M.A.; She, Y.; Shori, F.; Patmonoaji, A.; Matsushita, S.; Suekane, T.; Nagatsu, Y. Enhanced Heavy Oil Recovery by Calcium Hydroxide Flooding with the Production of Viscoelastic Materials: Study with 3-D X-Ray Tomography and 2-D Glass Micromodels. *Energy Fuels* **2021**, *35*, 11210–11222. [CrossRef]
30. Nagatsu, Y.; Abe, K.; Konmoto, K.; Omori, K. Chemical Flooding for Enhanced Heavy Oil Recovery via Chemical-Reaction-Producing Viscoelastic Material. *Energy Fuels* **2020**, *34*, 10655–10665. [CrossRef]
31. Coutinho, M.; Rocha, F.M.; Miranda, J.A. Viscous normal stresses and fingertip tripling in radial Hele-Shaw flows. *Phys. Rev. E* **2021**, *104*, 045106. [CrossRef]
32. Suzuki, R.X.; Kobayashi, S.; Nagatsu, Y.; Ban, T. Tunable Hydrodynamic Interfacial Instability by Controlling a Thermodynamic Parameter of Liquid–Liquid Phase Separation. *J. Phys. Chem. B* **2021**, *125*, 7508–7514. [CrossRef] [PubMed]
33. Stergiou, Y.; Eckert, K.; Schwarzenberger, K. Entrance effects in a radial Hele-Shaw cell: Numerical and experimental study. *Chem. Eng. J.* **2022**, *428*, 131146. [CrossRef]
34. Cieślik, A.; Akkermans, R.; Kamp, L.; Clercx, H.J.; van Heijst, G. Dipole-wall collision in a shallow fluid. *Eur. J. Mech.-B/Fluids* **2009**, *28*, 397–404. [CrossRef]
35. Lee, W.; Lucey, J. Structure and Physical Properties of Yogurt Gels: Effect of Inoculation Rate and Incubation Temperature. *J. Dairy Sci.* **2004**, *87*, 3153–3164. [CrossRef]
36. Bönisch, M.P.; Huss, M.; Lauber, S.; Kulozik, U. Yoghurt gel formation by means of enzymatic protein cross-linking during microbial fermentation. *Food Hydrocoll.* **2007**, *21*, 585–595. [CrossRef]
37. Wang, W.; Zhang, C.; Patmonoaji, A.; Hu, Y.; Matsushita, S.; Suekane, T.; Nagatsu, Y. Effect of gas generation by chemical reaction on viscous fingering in a Hele–Shaw cell. *Phys. Fluids* **2021**, *33*, 093104. [CrossRef]
38. Zittle, C. Precipitation of Casein from Acidic Solutions by Divalent Anions. *J. Dairy Sci.* **1966**, *49*, 361–364. [CrossRef]
39. Etrati, A.; Alba, K.; Frigaard, I.A. Two-layer displacement flow of miscible fluids with viscosity ratio: Experiments. *Phys. Fluids* **2018**, *30*, 052103. [CrossRef]
40. Chairunisa, W.; Imawan, C. The effect of pH on the characteristics of the methyl red solution as a gamma-ray dosimeter. *J. Phys. Conf. Ser.* **2019**, *1321*, 022015. [CrossRef]
41. Wang, L.; Nakanishi, Y.; Teston, A.D.; Suekane, T. Effect of diffusing layer thickness on the density-driven natural convection of miscible fluids in porous media: Modeling of mass transport. *J. Fluid Sci. Technol.* **2018**, *13*, JFST0002. [CrossRef]
42. Muhr, A.H.; Blanshard, J.M. Diffusion in gels. *Polymer* **1982**, *23*, 1012–1026. [CrossRef]
43. Patmonoaji, A.; Mahardika, M.A.; Nasir, M.; She, Y.; Wang, W.; Muflikhun, M.A.; Suekane, T. Stereolithography 3D Printer for Micromodel Fabrications with Comprehensive Accuracy Evaluation by Using Microtomography. *Geosciences* **2022**, *12*, 183. [CrossRef]

applied sciences

MDPI

Article

The Effect of Porous Media on Wave-Induced Sloshing in a Floating Tank

Wen-Huai Tsao [1,*], Ying-Chuan Chen [2], Christopher E. Kees [1] and Lance Manuel [3]

1 Department of Civil and Environmental Engineering, Louisiana State University, Baton Rouge, LA 70803, USA; cekees@lsu.edu

2 KBR, Houston, TX 77002, USA; ycchen@utexas.edu

3 Department of Civil, Architectural and Environmental Engineering, The University of Texas at Austin, Austin, TX 78712, USA; lmanuel@mail.utexas.edu

* Correspondence: whtsao@lsu.edu

Featured Application: This research can be applied to sloshing mitigation for LNG vessels and vibration control for offshore structures.

Abstract: Placing porous media in a water tank can change the dynamic characteristics of the sloshing fluid. Its extra damping effect can mitigate sloshing and, thereby, protect the integrity of a liquefied natural gas tank. In addition, the out-of-phase sloshing force enables the water tank to serve as a dynamic vibration absorber for floating structures in the ocean environment. The influence of porous media on wave-induced sloshing fluid in a floating tank and the associated interaction with the substructure in the ambient wave field are the focus of this study. The numerical coupling algorithm includes the potential-based Eulerian–Lagrangian method for fluid simulation and the Newmark time-integration method for rigid-body dynamics. An equivalent mechanical model for the sloshing fluid in a rectangular tank subject to pitch motion is proposed and validated. In this approach, the degrees of freedom modeling of the sloshing fluid can be reduced so the numerical computation is fast and inexpensive. The results of the linear mechanical model and the nonlinear Eulerian–Lagrangian method are correlated. The dynamic interaction between the sloshing fluid and floating body is characterized. The effectiveness of the added porous media in controlling the vibration and mitigating the sloshing response is confirmed through frequency response analysis.

Keywords: porous media; sloshing; equivalent mechanical model; fluid–structure interaction; floating platform

Citation: Tsao, W.-H.; Chen, Y.-C.; Kees, C.E.; Manuel, L. The Effect of Porous Media on Wave-Induced Sloshing in a Floating Tank. *Appl. Sci.* **2022**, *12*, 5587. https://doi.org/10.3390/app12115587

Academic Editor: Liangxing Li

Received: 6 May 2022
Accepted: 19 May 2022
Published: 31 May 2022

1. Introduction

Fluid–structure interaction (FSI) is the dynamic interaction between a deformable and moveable structure and its surrounding fluid. In recent years, this topic has been a major focus in civil and ocean engineering research because there has been an ongoing interest in exploring mineral and energy resources from the ocean. Various kinds of offshore work platforms have been developed for purposes of petroleum extraction, wind energy, wave energy, and fishing. As they are subject to external loadings from wind, waves, currents, ice, and earthquakes, their dynamics includes complex fluid–structure interactions. Very challenging engineering problems in this area include the seakeeping performance of a ship, as well as the longevity and integrity of the offshore structures arise. For Liquified Natural Gas (LNG) vessels, the rolling motion is of greatest concern because the ambient wave damping is usually too small to limit motion amplitudes, especially when resonance occurs. The strong excitations will induce violent sloshing behavior and high-speed waves in the liquid container. The extremely high impact on the wall of the tank will damage the inner metal membrane. Many researchers have studied the sloshing phenomena observed in ship tanks. Cariou and Casella [1] compared several typical numerical methods for sloshing analysis. They found that numerical computations for liquid impacts are very unpredictable. Kim et al. [2] used an efficient hybrid method that involves the impulse response function (IRF) for ship motions and a Reynold-averaged Navier–Stokes (RANS) solver for sloshing to simulate the sway motion of a barge with rectangular tanks and the roll motion of a hull with an anti-rolling tank (ART). Lee et al. [3] employed a similar approach for a floating production system equipped with two liquid tanks. They found that the coupling effect shifts the motion frequencies and induces the nonlinear motion amplitudes. Lyu et al. [4] compared the potential-flow and Navier–Stokes models. They concluded that the sloshing fluid can significantly influence the ship's surge and roll motions but hardly affects the heave motion. Their results showed that potential-based numerical methods are very suitable for most scenarios of sloshing in tanks on ships. To mitigate vibration, a common solution is to deploy bilge keels on the hull to increase the hydrodynamic resistance and damping from the vortex [5]. An ART, which consists of two vertical ducts connected at the bottom with a horizontal duct, distributed along the centerline of the ship, is another effective damper for rolling reduction. Such a liquid damper is also called a tuned liquid column damper (TLCD) in structural engineering. Their active control is enabled by installing propellers or pumps in the middle of the bottom tank [6–8]. For other floating structures, such as compliant towers, tension leg platforms, and floating production systems, the tuned mass damper (TMD), which is an auxiliary mass attached to the main structure through springs and dashpot elements, is a common vibration absorber [9,10]. It generates an out-of-phase control force against the external excitations and dissipates the kinetic energy of the main structure. Its performance can be maximized as the tuning condition is satisfied, i.e., the frequencies of the damper and structure are close. TMD systems have been widely applied to offshore wind turbines to handle some complicated phenomena such as gyroscopic effects, wind turbulence and shear, tower shadow, and the wake effect from neighboring turbines [11–13]. In the ocean environment, a tuned liquid damper (TLD), which is a water tank that dampens structural vibrations by the fluid viscosity, can be more suitable because of its convenience, economy, and simplicity [14,15]. However, its inherent viscosity is too small to provide sufficient damping and the accompanying nonlinear sloshing behavior can easily lead to mistuning [16]. The work of other researchers has sought to enhance the damping effect by placing internal obstructions such as baffles [17] and screens [18] inside the tank. Induced characteristic changes have been successively analyzed [19–21], but progress toward a practical application remains stagnant because it is difficult to satisfy the tuning condition and optimal damping at the same time. Another new type of TLD with porous media was proposed to seek optimal performance through a mechanical model [22,23]. Tsao and Chang [24] carried out free decay experiments to verify the linear dynamic characteristics. Tsao et al. [25] calculated the nonlinear damping ratio using resonance experiments. Tsao and Huang [26] considered the Forchheimer flow in the numerical model and derived two essential parameters that appear in the mathematical model. However, the previous research studies only focused on the characteristics of the TLD systems instead of their application in FSI problems.

For FSI modeling, the key point is the coupling algorithm that connects the numerical methods for the fluid and rigid-body dynamics to ensure kinematic continuity (same displacement and velocity) and dynamic continuity (same pressure) on the fluid–solid interface. Many researchers considered potential-flow theory. Sen [27] obtained the hydrodynamic force equilibrium of the body motion iteratively. Van Daalen [28] combined the Bernoulli equation and the motion equation of the body to derive an additional integral equation that satisfies the continuity simultaneously. His method requires little extra computational effort but avoids numerical iterations. Tanizawa [29] provided a similar formula in the implicit algorithm. Jung et al. [30] studied the viscous effect around a rolling floating body through wave tank experiments. They concluded that eddy-making damping is important to calculate the roll motion. Guerber et al. [31] and Dombre et al. [32] compared several numerical schemes for the forced and free motions of a fully submerged body.

Other useful methods for solving the FSI problem in potential flow can be observed in past articles [33–35]. In recent decades, more researchers study FSI problems via RANS analysis. Compared to the potential-flow model, it can describe the fluid reaction to structures more appropriately by including viscosity, the boundary layer, vortex, and breaking waves. Matthies et al. [36] applied the Arbitrary Lagrangian–Eulerian (ALE) framework to build a strongly coupled algorithm. Palm et al. [37] used the volume-of-fluid method to capture the motion of the wave energy converter. Tezduyar [38] demonstrated the numerical results of some typical problems in ocean engineering, such as propellers, pipes, free-surface flow, and the sloshing phenomenon. Rakhsha et al. [39] compared the Lagrangian and Eulerian approaches (SPH vs. FEM) in FSI simulations. They emphasized the numerical robustness, ease of model setup, and versatility for several typical scenarios including the floating and submerged bodies and the elastic gate of a dam. More numerical approaches and their applications can be observed in textbooks [40].

This paper emphasizes the influence of porous media on the interaction between wave-induced sloshing in a tank and the dynamics of the attached floating body. The application of porous media to structural vibration control and sloshing mitigation will be addressed as well. The Eulerian–Lagrangian method was adopted for the hydrodynamic computation in both the pore-flow domain (sloshing fluid in porous media) and the pure-water domain (ambient waves). This numerical method includes the mixed-type boundary value problem (BVP) solver implemented by the boundary element method (BEM) and the free-surface particle tracker via the second-order Taylor series expansion. Similar approaches had been applied to solve wave absorption [41], shoaling [42], and the local impact of a water jet [43]. Another equivalent mechanical model was employed to calculate the sloshing responses. This mechanical model involves several sets of mass-spring-dashpot systems, hence the degree of freedom is greatly reduced and the fluid computation is replaced by a few motion equations. The nonlinear damping of the mechanical model was validated via shaking table tests [25] but the results were obtained only when the tank was subject to horizontal excitations. Therefore, the supplementary solution for the mechanical model undergoing angular excitation is provided in this paper. For the rigid-body dynamics, the Newmark time-integration method was employed to ensure numerical stability. The body dynamics can be solved as long as the hydrodynamic resultants are given. The hydrodynamic forces are not, however, known a priori; they can only be obtained by solving the BVP that involves the body dynamics as the boundary conditions. In this work, therefore, an iterative algorithm was adopted to ensure the equilibrium between the body motion and fluid reaction, i.e., Newton's second law of motion. Section 2 describes the details of the numerical methods used for the fluid and rigid-body dynamics and the coupling algorithm. Section 3 presents the dynamic characteristics of the floating body. Section 4 explains the FSI between the sloshing fluid in porous media, floating body, and waves via several benchmark tests. The conclusions are addressed in Section 5.

2. Methodology

In this section, the governing equations and boundary conditions for the pore-flow domain and pure-water domain are addressed. We adopt an Eulerian–Lagrangian (EL) method, which solves both flow fields in a similar framework. The equivalent mechanical model is provided only for the sloshing phenomenon in porous media. The time-integration method for rigid-body dynamics and the iterative algorithm for strong fluid–solid coupling are presented.

2.1. Governing Equations and Boundary Conditions

A two-dimensional rectangular water tank filled with porous media on the rectangular floating platform in waves is shown in Figure 1a. Assuming that the fluid is incompressible, inviscid, and irrotational, the continuity and momentum equations of the fluid can be expressed as:

$$\nabla \cdot \boldsymbol{u} = 0 \tag{1}$$

$$\rho\left[\frac{\partial \boldsymbol{u}}{\partial t}+(\boldsymbol{u}\cdot\nabla)\boldsymbol{u}\right]=-\nabla p+\nabla p_d+\rho \boldsymbol{g} \tag{2}$$

where ρ is the fluid density, $\boldsymbol{u}$ is the fluid velocity, p is the pressure, p_d is the pressure drop due to the porous media, and $\boldsymbol{g}$ is the gravitational acceleration. Note that in fresh water, $\nabla p_d = 0$, and nonlinearity arises only from the convective term; when the porous media is involved, the gradient of the pressure drop can be expressed by [44]:

$$\nabla p_d=-\frac{\gamma\mu}{\kappa}(\boldsymbol{u}-\boldsymbol{v})-\frac{\gamma c_F\rho}{\sqrt{\kappa}}|\boldsymbol{u}-\boldsymbol{v}|(\boldsymbol{u}-\boldsymbol{v}) \tag{3}$$

where $\boldsymbol{v}$ is the velocity of the porous media, μ is the dynamic viscosity of the fluid, γ and κ are the porosity and permeability of the porous media, respectively, and c_F is a dimensionless constant determined by experiments. Note that if $\boldsymbol{u}$ is sufficiently small, the first linear term on the right-hand side of Equation (3) dominates, while the second quadratic function becomes significant as $\boldsymbol{u}$ increases and the drag due to solid obstacles is comparable with the surface drag due to friction. In potential flow, the fluid velocity can be expressed by the gradient of the potential function ϕ as:

$$\boldsymbol{u}=\nabla\phi \tag{4}$$

Therefore, the Laplace equation should be satisfied as:

$$\nabla^2\phi=0 \tag{5}$$

The kinematic boundary condition on the free surface can be expressed as:

$$\frac{D\boldsymbol{R}}{Dt}=\boldsymbol{u} \tag{6}$$

where $\boldsymbol{R}$ is the location of a free-surface particle. For the wave channel, the dynamic boundary condition on the free surface can be expressed as:

$$\frac{\partial\phi}{\partial t}+\frac{1}{2}|\nabla\phi|^2+g\eta=0 \tag{7}$$

where η is the wave elevation. For the sloshing fluid, the dynamic boundary condition on the free surface can be expressed as:

$$\frac{\partial\phi}{\partial t}+\frac{1}{2}|\nabla\phi|^2+(\phi-\psi)(\alpha_1+\alpha_2|\nabla\phi-\nabla\psi|)+g\eta=0 \tag{8}$$

where $\alpha_1=\gamma\mu/\kappa\rho$ and $\alpha_2=c_F/\sqrt{\kappa}$ are constants related to the nature of the porous media and ψ is the velocity potential of the tank. On the wetted boundaries, the impermeable condition is taken as:

$$\frac{\partial\phi}{\partial n}=\dot{\boldsymbol{x}}\cdot\boldsymbol{n} \tag{9}$$

where $\boldsymbol{n}$ denotes the normal vector pointing outward to the fluid boundary and $\dot{\boldsymbol{x}}$ is the velocity of the node on the wetted boundary. As a wave propagates along the channel, the radiation condition is applied on the outlet boundary to ensure the waves scatter to infinity as [41]:

$$\frac{\partial\phi}{\partial n}=-\frac{1}{c}\frac{\partial\phi}{\partial t} \tag{10}$$

where c is the wave velocity.

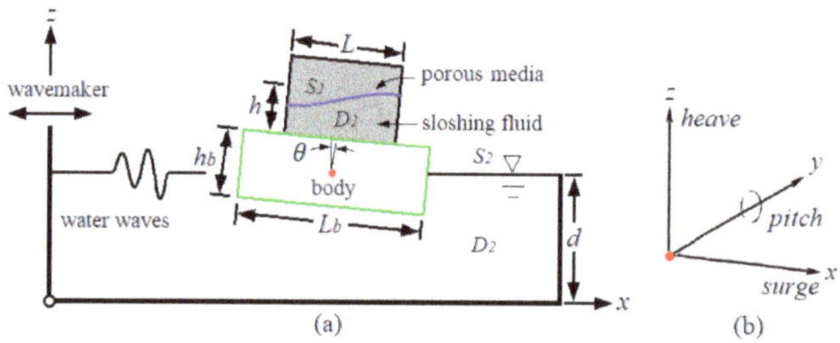

Figure 1. (**a**) Configuration of a floating body undergoing wave excitations with a water tank, installed with porous media, attached above; (**b**) degree of freedom of the floating body motion.

2.2. Eulerian–Lagrangian Method

By applying Green's second identity to the Laplace equation, the boundary integral equation (BIE) for the fluid boundary can be obtained as:

$$c_p\phi_p = \int_S \left(\frac{\partial G}{\partial n}\phi - G\frac{\partial \phi}{\partial n} \right) dS \tag{11}$$

where G is the fundamental solution with source point p and c_p is the resulting flux. In this paper, BIE is discretized, and the associated integrals are implemented by BEM. Therefore, Equation (11) can be rewritten in matrix form as:

$$\mathbf{A}\{\phi\} = \mathbf{B}\{\phi_n\} \tag{12}$$

where $\mathbf{A}$ and $\mathbf{B}$ are the kernel matrices and $\{\phi\}$ and $\{\phi_n\}$ are the vectors containing the potential and the normal velocity, respectively, at all of the boundary nodes. Equation (12) represents a mixed-type BVP with ϕ given on the free surface and ϕ_n on the wetted boundaries. The free-surface particle is tracked via a second-order Taylor series expansion, assuming a small time step, Δt, as follows:

$$\boldsymbol{R}(t+\Delta t) = \boldsymbol{R}(t) + \Delta t\frac{D\boldsymbol{R}}{Dt} + \frac{\Delta t^2}{2}\frac{D^2\boldsymbol{R}}{Dt^2} + O\left(\Delta t^3\right) \tag{13}$$

$$\phi(t+\Delta t) = \phi(t) + \Delta t\frac{D\phi}{Dt} + \frac{\Delta t^2}{2}\frac{D^2\phi}{Dt^2} + O\left(\Delta t^3\right) \tag{14}$$

The first-order derivative in Equation (13) is obtained by the kinematic boundary condition, while the first-order derivative in Equation (14) can be obtained by the momentum equations, i.e., Equation (7) for water waves and Equation (8) for sloshing water. The second-order derivatives of $\boldsymbol{R}$ and ϕ can be obtained as:

$$\frac{D^2R_i}{Dt^2} = \frac{\partial^2\phi}{\partial t\partial \boldsymbol{x}} + \boldsymbol{u}\nabla\cdot\frac{\partial\phi}{\partial \boldsymbol{x}_i} \tag{15}$$

$$\frac{D^2\phi}{Dt^2} = \boldsymbol{u}\cdot\frac{D\boldsymbol{u}}{Dt} - g\frac{D\eta}{Dt} \tag{16}$$

A new unknown variable $\phi_t = \partial\phi/\partial t$ appears in the equations above. Since ϕ_t also satisfies the Laplace equation, another mixed-type BVP involving ϕ_t and ϕ_{tn} arises and needs to be solved. The Dirichlet condition on the free surface is given by:

$$\phi_t = \frac{D\phi}{Dt} - |\nabla\phi|^2 \tag{17}$$

The Neumann condition on the wetted boundaries is given by:

$$\phi_{tn} = \ddot{\boldsymbol{x}}\cdot\boldsymbol{n} - (\boldsymbol{u}\cdot\nabla)\phi_n, \tag{18}$$

while the Neumann condition on the radiation boundary is given by:

$$\phi_{tn} = c\frac{\partial^2 \phi}{\partial s^2} \tag{19}$$

where $\ddot{x}$ is the acceleration of the node on the wetted boundary and s denotes the tangential direction of the boundary. Fortunately, the domain geometry remains unchanged at the current time step, so the matrices **A** and **B** in Equation (12) can be reused.

2.3. Floating Body Dynamics and FSI Coupling

In this paper, the floating body motion has three degrees of freedom, namely it is free to surge (horizontal movement along the x direction), heave (vertical movement along the z direction), and pitch (angular movement about the y-axis), as shown in Figure 1b. Therefore, the equations of translational and angular motions can be expressed as:

$$M_b\ddot{\boldsymbol{x}}_G = \boldsymbol{F} + M_b\boldsymbol{g} \tag{20}$$

$$I_G\ddot{\boldsymbol{\theta}}_G = \boldsymbol{M} \tag{21}$$

where $\ddot{\boldsymbol{x}}_G$ and $\ddot{\boldsymbol{\theta}}_G$ are the translational and angular acceleration of the body center, respectively, while M_b and I_G are the mass and moment of inertia of the body, respectively. The resultant force $\boldsymbol{F}$ and overturning moment $\boldsymbol{M}$ arising from the sloshing fluid and waves can be expressed as:

$$\boldsymbol{F} = \int_{S_{w1}+S_{w2}} p\boldsymbol{n}\, dS \tag{22}$$

$$\boldsymbol{M} = \int_{S_{w1}+S_{w2}} p(\boldsymbol{r}\times\boldsymbol{n})\, dS \tag{23}$$

where S_{w1} and S_{w2} are the wetted boundaries of the floating body's contact with the sloshing water and waves, respectively, and $\boldsymbol{r}$ is the position vector of a fluid particle with respect to the body center. The pressure can be obtained by the Bernoulli equation as:

$$p = -\rho\left(\frac{\partial\phi}{\partial t} + \frac{1}{2}|\nabla\phi|^2 + g\eta\right) \tag{24}$$

To solve the motion equations, the Newmark method based on the constant average acceleration assumption is employed as:

$$\ddot{\boldsymbol{x}}_{G,k+1} = \frac{\boldsymbol{F}_{k+1}}{M_b} \tag{25}$$

$$\dot{\boldsymbol{x}}_{G,k+1} = \dot{\boldsymbol{x}}_{G,k} + 0.5\Delta t\left(\ddot{\boldsymbol{x}}_{G,k} + \ddot{\boldsymbol{x}}_{G,k+1}\right) \tag{26}$$

$$\boldsymbol{x}_{G,k+1} = \boldsymbol{x}_{G,k} + \Delta t\dot{\boldsymbol{x}}_{G,k} + 0.25\Delta t^2\left(\ddot{\boldsymbol{x}}_{G,k} + \ddot{\boldsymbol{x}}_{G,k+1}\right) \tag{27}$$

where k is the time step. Note that the above equations are available for solving the translational motion, while the angular motion can be solved by switching the force and mass in Equation (25) to the overturning moment and moment of inertia, respectively. This scheme is unconditionally stable and it does not need a sub-step. However, the resultant force is not known a priori because, to solve the fluid field, the body kinematics used in Equations (9) and (18) should be given but they depend on the body motion as:

$$\dot{x} = \left(\dot{\boldsymbol{x}}_G + \boldsymbol{\theta}_G \times \boldsymbol{r}\right)\cdot\boldsymbol{n} \tag{28}$$

$$\ddot{x} = \left[\ddot{\boldsymbol{x}}_G + \ddot{\boldsymbol{\theta}}_G \times \boldsymbol{r} + \dot{\boldsymbol{\theta}}_G \times \left(\dot{\boldsymbol{\theta}}_G \times \boldsymbol{r}\right)\right]\cdot\boldsymbol{n} \tag{29}$$

Therefore, an iterative algorithm, illustrated in Algorithm 1, is used to ensure equilibrium between the body motion and fluid reaction.

Algorithm 1: An iterative algorithm for FSI by EL method.

```
do k > 0
  Start iterative step with the solution at kth time step as an initial guess
  ε = abs(ẍ_{G,k,i} − ẍ_{G,k,i−1})
  do while ε > ε_tolerance
  Solve φ_i and φ_{n,i} (the first BVP)
  Solve φ_{t,i} and φ_{tn,i} (the second BVP)
  Calculate pressure along wetted boundaries
  Calculate force and moment act on the floating body
  Define predictor of force and moment: F* = F_{k,i} + ω(F*_{i−1} − F_{k,i})
  Solve rigid-body motion (Newmark method)
  Update free surface and body boundary S_{k,i+1} and φ_{k,i+1}
  end do
  x_{k+1} = x_{k,i+1}
  ẋ_{k+1} = ẋ_{k,i+1}
  ẍ_{k+1} = ẍ_{k,i+1}
  φ_{k+1} = φ_{k,i+1}
  S_{k+1} = S_{k,i+1}
end do
```

2.4. Equivalent Mechanical Model

In addition to the EL method, the mechanical model can be useful to solve the sloshing problem in porous media by means of a series of motion equations. For the rectangular tank, the mechanical model consists of a fixed mass m_f located at the height z_f, and infinite mass-dashpot-spring systems (denoted by m_n, c_n, and k_n) located at the heights z_n as shown in Figure 2. The equivalent masses, stiffness, damping ratios, and location heights can be expressed as [24]:

$$m_n = \frac{8\gamma\rho B}{a_n^3 L} tanh\, a_n h \tag{30}$$

$$k_n = m_n\omega_n^2 = \frac{8\gamma\rho g B}{a_n^2 L} tanh^2 a_n h \tag{31}$$

$$\xi_n = \frac{\alpha_1}{2\omega_n} = \frac{\gamma\mu}{2\kappa\rho\omega_n} = \frac{\gamma\mu}{2\kappa\rho}\sqrt{\frac{cotha_n h}{a_n g}} \tag{32}$$

$$z_n = h + \frac{2}{a_n sinha_n h} - \frac{1}{a_n tanha_n h} \tag{33}$$

$$m_f = m_w - \sum\nolimits_{n=0}^{\infty} m_n = \gamma\rho BLh\left[1 - \sum_{n=0}^{\infty}\frac{8}{a_n^3 L^2 h} tanha_n h\right] \tag{34}$$

$$z_f = \frac{1}{m_f}\left[\gamma\rho B\left(\frac{Lh^2}{2} + \frac{L^3}{12}\right) - \sum_{n=0}^{\infty} m_n z_n\right] \tag{35}$$

where $a_n = (2n+1)\pi/L$, $n = 0, 1, 2\cdots$, $m_w = \gamma\rho BLh$ denotes the total water mass in the tank, and B is the tank width. The above equivalent parameters are obtained when the tank is subject to horizontal excitations. However, it is not sufficient to describe the angular motion studied in this paper. The moment of inertia should be appended to satisfy dynamic equilibrium. As the tank is subjected to a harmonic pitch motion, its angular displacement is expressed by:

$$\theta(t) = \theta_0 sin\omega t \tag{36}$$

where θ_0 is the amplitude of angular displacement. The velocity potential on the walls and bottom should satisfy the impermeable boundary condition as:

$$\frac{\partial\phi}{\partial x} = \theta_0\omega z cos\omega t, \text{ on wetted walls} \tag{37}$$

$$\frac{\partial \phi}{\partial z} = -\theta_0 \omega x cos\omega t, \text{ on bottom} \tag{38}$$

Hence the velocity potential can be expressed as:

$$\begin{aligned}\phi = \theta_0 \omega cos\omega t \quad & \sum_{n=0}^{\infty} 4(-1)^n \left[\frac{sinb_n\left(z-\frac{h}{2}\right)}{b_n^3 h} \frac{sinhb_n x}{coshb_n \frac{L}{2}} + \frac{sina_n x}{a_n^3 L} \frac{cosha_n(z-h)}{sinha_n h} \right] \\ & + \theta_0 \omega \sum_{n=0}^{\infty} (D_{1n} cos\omega t + D_{2n} sin\omega t) sina_n x \, cosha_n z\end{aligned} \tag{39}$$

where $b_n = (2n+1)\pi/h$. On the free surface, the linear kinematic and dynamic boundary conditions are combined and applied at $z = h$ as:

$$\frac{\partial^2 \phi}{\partial t^2} + \alpha_1 \left(\frac{\partial \phi}{\partial t} - \frac{\partial \psi}{\partial t} \right) + g \frac{\partial \phi}{\partial z} = 0 \tag{40}$$

Substituting Equation (39) into Equation (40), the coefficients D_{1n} and D_{2n} can be solved as:

$$D_{1n} = \frac{-p_n}{p_n^2 + q_n^2} \frac{4\omega^2}{sina_n x cosha_n h} \left[\frac{sinhb_n x}{b_n^3 h coshb_n \frac{L}{2}} + \frac{(-1)^n sina_n x}{a_n^3 L sinha_n h} \right] \tag{41}$$

$$D_{2n} = \frac{q_n}{p_n^2 + q_n^2} \frac{4\omega^2}{sina_n x cosha_n h} \left[\frac{sinhb_n x}{b_n^3 h coshb_n \frac{L}{2}} + \frac{(-1)^n sina_n x}{a_n^3 L sinha_n h} \right] \tag{42}$$

where $p_n = \omega^2 - g a_n tanha_n h$ and $q_n = \gamma \alpha_1 \omega$. The overturning moment due to the sloshing fluid can be obtained by integrating the pressure along the wetted boundary as:

$$\begin{aligned} M_o = \gamma\rho \int_0^h \int_0^B \left(\left.\frac{\partial \phi}{\partial t}\right|_{x=\frac{L}{2}} - \left.\frac{\partial \phi}{\partial t}\right|_{x=\frac{-L}{2}} \right) zdydz + \gamma\rho \int_{-L/2}^{L/2} \int_0^B \left.\frac{\partial \phi}{\partial t}\right|_{z=0} xdydx \\ = -\gamma\rho B \theta_0 \omega^2 sin\omega t \sum_{n=0}^{\infty} 8 \left[\frac{3sinhb_n \frac{L}{2} - b_n \frac{L}{2} coshb_n \frac{L}{2}}{b_n^5 h coshb_n \frac{L}{2}} + \frac{2cosha_n h - 1}{a_n^5 L sinha_n h} \right] \\ + \gamma\rho B \theta_0 \omega^2 \sum_{n=0}^{\infty} \frac{8\omega^2 (p_n sin\omega t + q_n cos\omega t)}{(p_n^2 + q_n^2) cosha_n h} \left[\frac{b_n \frac{L}{2} coshb_n \frac{L}{2} - sinhb_n \frac{L}{2}}{b_n^5 h coshb_n \frac{L}{2}} \right. \\ \left. + \frac{a_n h sinha_n h - cosha_n h + 2}{a_n^5 L sinha_n h} + \frac{(a_n h sinha_n h - cosha_n h + 1) tanhb_n \frac{L}{2}}{a_n^2 b_n^3 h} \right] \end{aligned} \tag{43}$$

Based on the dynamic equivalent, the mechanical model should give the same overturning moment under the same circumstances. Therefore, the motion equation of the mechanical system can be written as:

$$m_n \ddot{x}_n + c_n \dot{x}_n + k_n x_n = m_n z_n \theta_0 \omega^2 sin\omega t \tag{44}$$

where x_n is the equivalent displacement of the moving mass relative to the ground. The overturning moment can be obtained as:

$$\begin{aligned} \overline{M_o} &= I_e \ddot{\theta} + m_f z_f z_f \ddot{\theta} + \sum_{n=0}^{\infty} m_n z_n z_n \ddot{\theta} + \sum_{n=0}^{\infty} m_n z_n \ddot{x}_n \\ &= \left(I_e + m_f z_f^2 + \sum_{n=0}^{\infty} m_n z_n^2 \right) \ddot{\theta} + \sum_{n=0}^{\infty} m_n z_n \ddot{x}_n \end{aligned} \tag{45}$$

Using the same technique described in reference [24] to equalize the overturning moments due to rigid-body motion (regardless of the oscillatory component), i.e., the first term on the right-hand side of Equations (43) and (45), the moment of inertia I_e can be obtained as:

$$I_e = 8\gamma\rho B \sum_{n=0}^{\infty} \left(\frac{3sinhb_n\frac{L}{2} - b_n\frac{L}{2}coshb_n\frac{L}{2}}{b_n^5 hcoshb_n\frac{L}{2}} + \frac{2cosha_n h - 1}{a_n^5 Lsinha_n h} - m_n z_n^2 \right) - m_f z_f^2 \tag{46}$$

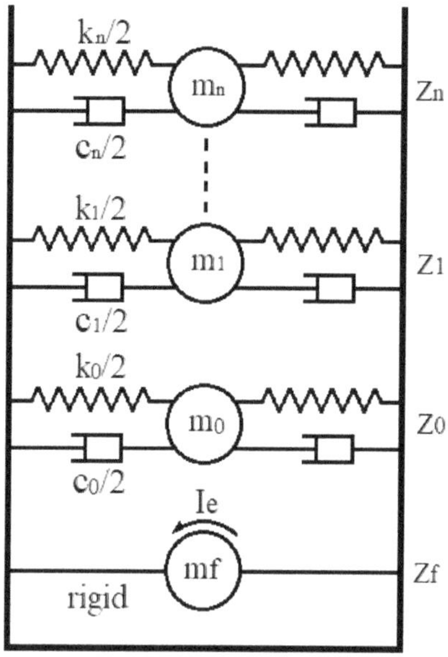

Figure 2. The schematic diagram of the equivalent mechanical model of a water tank filled with porous media.

When the mechanical model is applied to the floating body, its motion equation can be expressed as:

$$m_n\ddot{x}_n + c_n\dot{x}_n + k_n x_n = -m_n\ddot{x}_G + m_n z_n\ddot{\theta}_G \tag{47}$$

Therefore, Equation (25) for the surge motion of the floating body can be modified by:

$$M_b\ddot{x}_G = \int_{S_{w2}} pn_x dS + m_f z_f\ddot{\theta}_G + \sum_{n=0}^{\infty}\left(c_n\dot{x}_n + k_n x_n + m_n z_n\ddot{\theta}_G\right) \tag{48}$$

For the heave motion, the mechanical model assumes no masses move vertically relative to the tank, so the entire system will be regarded as a dead load to the floating body. Therefore, the equilibrium equation can be simplified as:

$$M_b\ddot{z}_G = \int_{S_{w2}} pn_z dS - m_w g \tag{49}$$

For the pitch motion, the movement of the equivalent masses will produce an external moment on the floating body; therefore, the equilibrium equation can be modified by:

$$I_b\ddot{\theta}_G = \int_{S_{w2}} p(\boldsymbol{r}\times\boldsymbol{n})\, dS - \left(I_e + m_f z_f^2\right)\ddot{\theta}_G - \sum_{n=0}^{\infty}\left(c_n\dot{x}_n + k_n x_n + m_n z_n\ddot{\theta}_G\right)z_n \tag{50}$$

During the computation of Equations (48)–(50), the resultants from the sloshing behavior can be obtained once Equation (47) is solved. Since the sloshing computation does not go through the EL method, the numerical simulation is very efficient and inexpensive.

3. Characteristics Analysis for Floating Body

The dynamic characteristics of a rectangular floating body were examined through a free-vibration test. The channel length = 10 m and water depth = 2 m. The rectangular floating platform had its length = 1 m, height = 0.5 m, and density = 450 kg/m^3, meaning the body mass = 225 kg and moment of inertia = 23.44 kg-m^2. Its center of gravity was placed at x = 5m and y = 2m, which was at the same height as the still water surface. The initial inclined angle was 5°. A total of 1714 two-node linear elements were distributed on the boundary of the water in the channel. The time interval was 0.001 s. As no incident waves were applied, the free vibration of the surge, heave, and pitch motions of the body are shown in Figures 3–5, respectively. The corresponding response spectra are shown in Figure 6. In Figure 3, the horizontal drift was small, even without the mooring constraint. Since the horizontal force came from the pressure difference between the left and right sides of the body, which was in sync with the pitch motion, the surge velocity and acceleration had the same oscillatory frequency of 0.58 Hz as the pitch motion, as shown in Figure 6. The heaving frequency = 0.66 Hz. Compared to the solution of a hanging spring model, regardless of the added mass effect (0.74 Hz), the numerical result was reasonable for a partially immersed body. In Figure 4, the heave motion decayed because the storing potential energy was passed to ambient waves and then carried away through the radiation condition. It eventually reached the static displacement of 0.0245 m. For the pitch motion, slight decay was observed. This has been explained in the past article [30]; the amplitude of the pitch motion could be over-predicted in potential flow since the wave damping was neglected. However, the pitching frequency = 0.58 Hz would not be affected. All the body motions were quite linear.

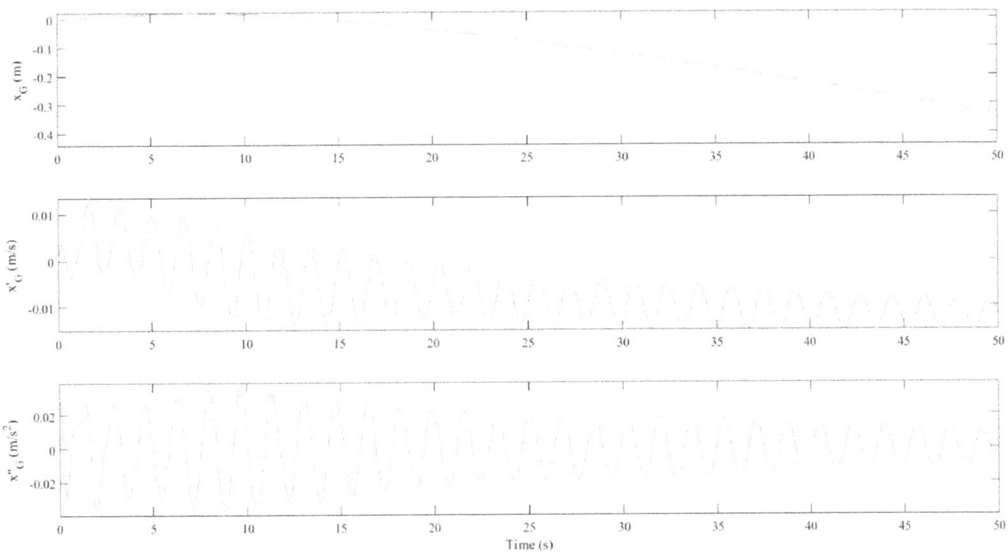

Figure 3. The horizontal displacement, velocity, and acceleration time histories of the floating body under free vibration at an initial 5° pitch.

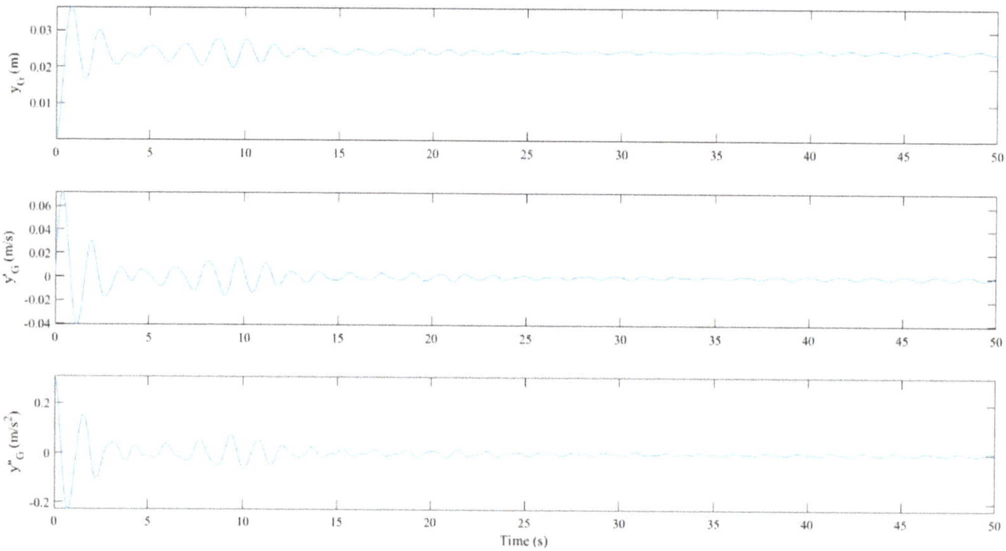

Figure 4. The vertical displacement, velocity, and acceleration time histories of the floating body under free vibration at an initial 5° pitch.

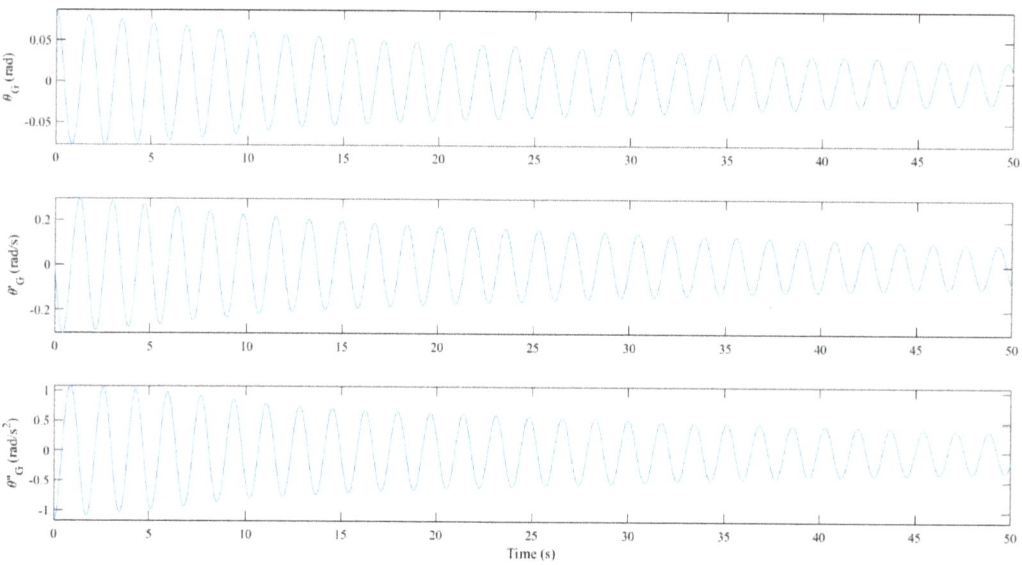

Figure 5. The angular displacement, velocity, and acceleration time histories of the floating body under free vibration at an initial 5° pitch.

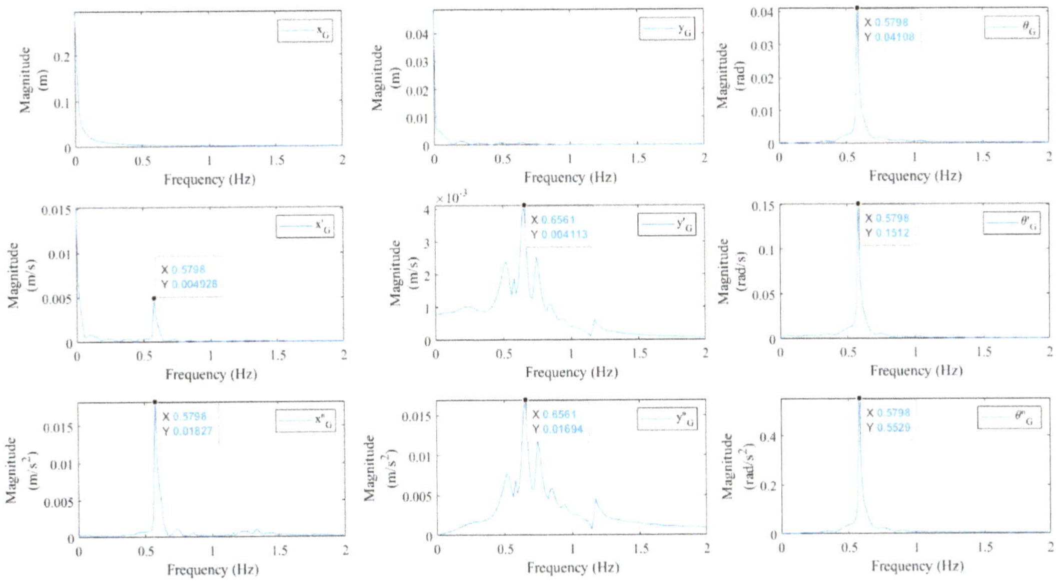

Figure 6. The response spectra of the floating body under free vibration at an initial 5° pitch.

4. Results and Discussion

The first example compares the nonlinear EL method and the linear mechanical model. The accuracy of the mechanical model on pitch-motion simulation is verified. The second example presents the numerical results of a water tank with and without porous media to accentuate the effect of porous media on the wave-induced sloshing phenomenon. The third example shows the benefit of this research in vibration control for offshore structures and sloshing mitigation for LNG vessels via frequency response analysis.

4.1. Sloshing in a Tank Filled with Porous Media: EL Method vs. Mechanical Model

In this example, the configuration of a rectangular body in the wave channel is the same as described in Section 3. The piston-type wavemaker was set on the left side of the channel to generate harmonic waves that propagated along the x direction to the right side. The wave frequency was 0.58 Hz, which was resonant with the floating body. Two cases were tested. The large wave had its wave height = 0.05 m, so the demand stroke of the wavemaker was 2.64 cm, while the small wave had its wave height = 0.03 m, so the demand stroke of the wavemaker was 1.58 cm [45]. The water tank was installed in a way in which the bottom was aligned with the centerline of the floating body. The tank length was 0.64 m and the water depth was 0.047 m, hence the fundamental sloshing frequency was 0.53 Hz. The masses of the tank and porous media were neglected. The porosity of the porous media = 0.96. Assume the first modal damping ratio was 18.4%, which results in α_1 = 1.28 s^{-1} and α_2 = 5.0 m^{-1} being used in the EL method [26]. A total of 80 two-node linear elements were distributed on the boundary of the sloshing fluid. When the mechanical model was applied, both the single-degree-of-freedom (SDOF) (namely, only the fundamental sloshing mode was considered) and five-degree-of-freedom (5-DOF) (namely, the first five sloshing modes were considered) systems were tested. The numerical configurations for the EL method and mechanical model are shown in Figure 7.

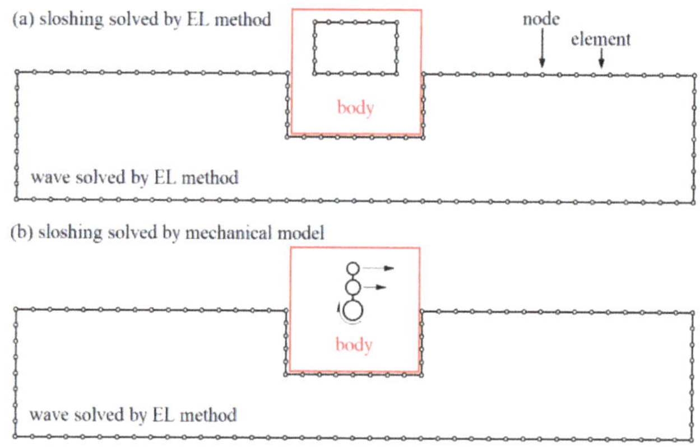

Figure 7. The numerical configurations for (**a**) EL method; (**b**) mechanical model, in the FSI problem (not to scale).

The surge motion responses of the floating body undergoing large wave excitation are shown in Figure 8. The body drifted along the x direction due to the continuous incoming waves. The numerical displacements were slightly different but the accelerations and velocities were very close. Since the dynamic responses depended on the external forces, their correlation will be explained via the horizontal forces as shown in Figure 9. The sloshing force cancels out the wave load due to phase cancellation. The EL method had a larger sloshing reaction, hence a smaller total force and a smaller acceleration were obtained. It is noticeable that in past experiments [25], the mechanical model overestimated the sloshing force. However, in our case, the external excitations to the water tank were different as the coupling effect was involved. Therefore, one may not always expect an over-predicted value from the mechanical model.

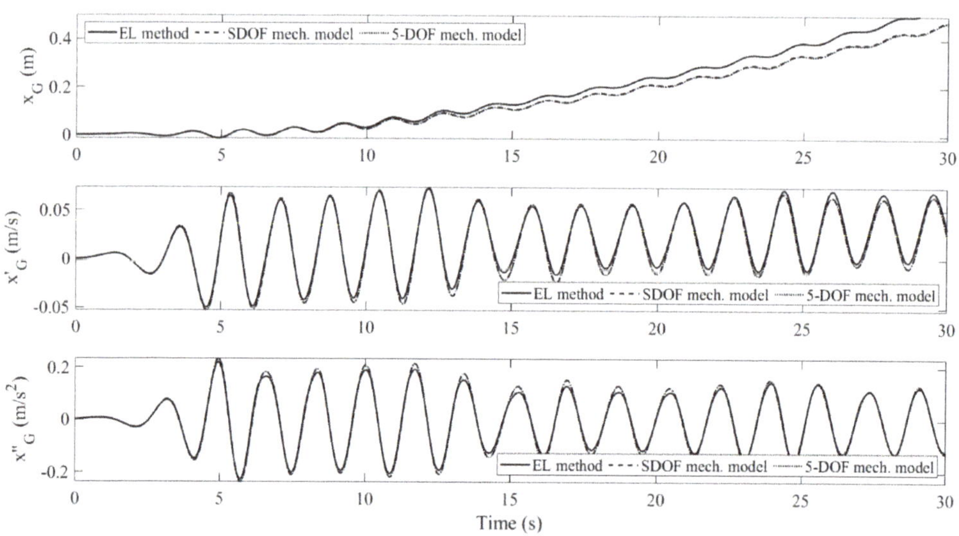

Figure 8. The surge motion responses of the floating body, configured with a water tank installed with porous media, undergoing large wave excitation.

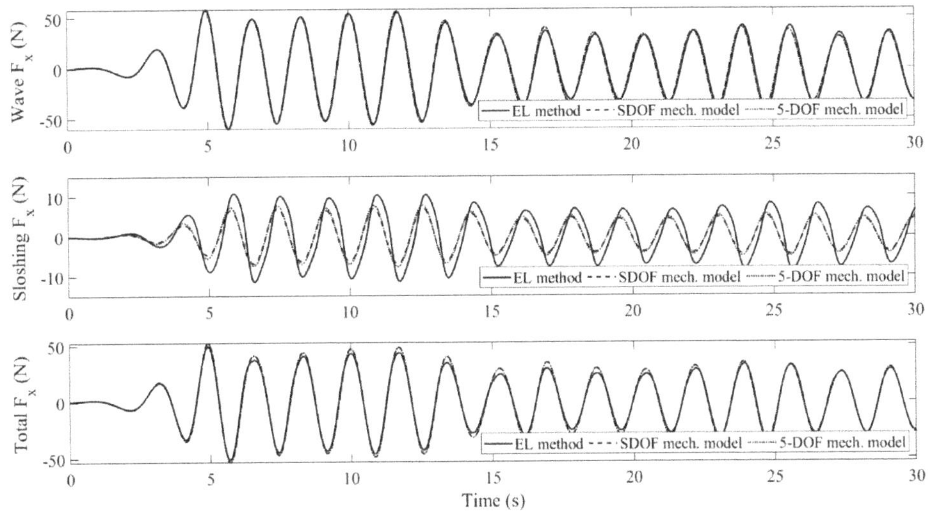

Figure 9. The horizontal forces acting on the floating body for the large-wave case.

The heave motion responses of the floating body and the vertical resultant forces to the body are shown in Figures 10 and 11, respectively. The mean value of the steady-state displacement was −0.0043 m, which was the same as the static displacement. The responses by both methods were identical except for the sloshing force. The mechanical model assumes the vertical sloshing force as a dead load, while the EL method takes the dynamic force due to the momentum change and convective acceleration into account. However, this effect was not significant in the vertical direction. It only caused a 3% (10 N) deviation from the dead load. Therefore, the heave response remained the same.

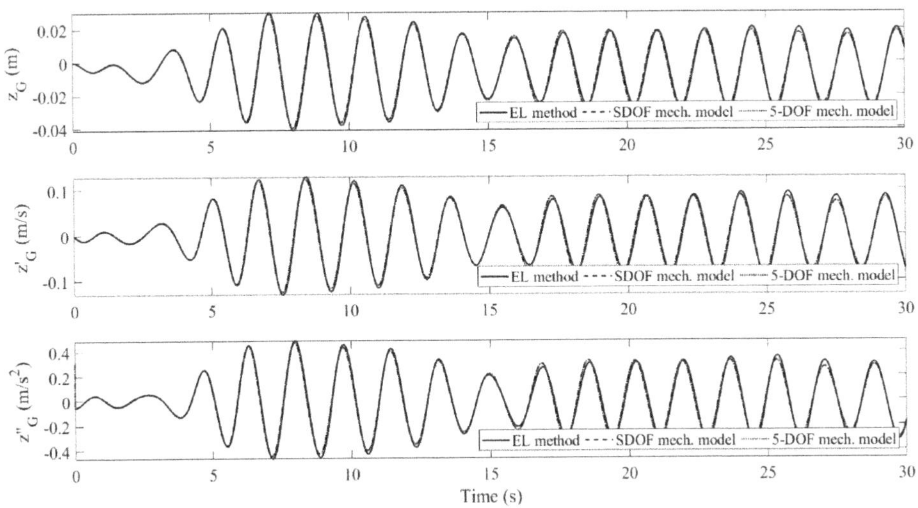

Figure 10. The heave motion responses of the floating body, configured with a water tank installed with porous media, undergoing large wave excitation.

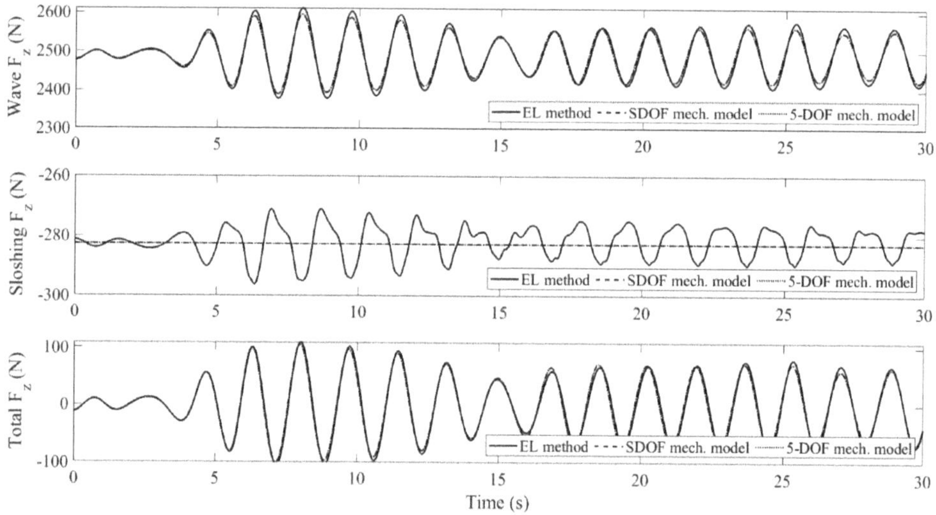

Figure 11. The vertical forces acting on the floating body for the large-wave case.

The pitch motion responses of the floating body and the overturning moments to the body are shown in Figures 12 and 13, respectively. The sloshing overturning moment by the EL method was nonlinear, while that by the mechanical model was linear. It is important to take the nonlinear effect into account since it changed the pitch amplitudes of the body in this example. The phase cancellation appeared. Since the sloshing-induced moment was comparable to the wave-induced moment, the moment cancellation was fairly strong. The body responses and sloshing reactions by the SDOF and five-DOF mechanical systems were identical for the surge, heave, and pitch motions. This shows that the fundamental mode dominated and the high-mode sloshing responses were too small to affect the FSI behavior. It justifies the design strategies for the liquid dampers that only emphasized the first modal response [22].

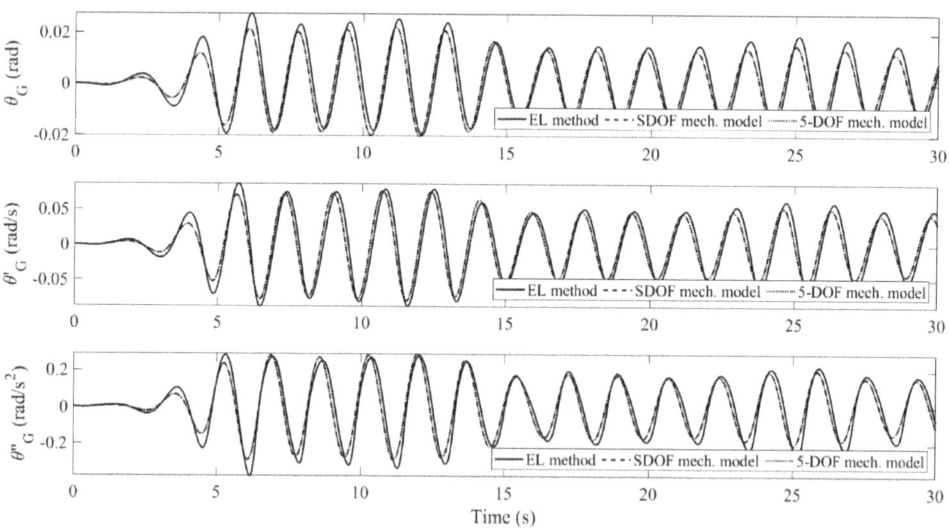

Figure 12. The pitch motion responses of the floating body, configured with a water tank installed with porous media, undergoing large wave excitation.

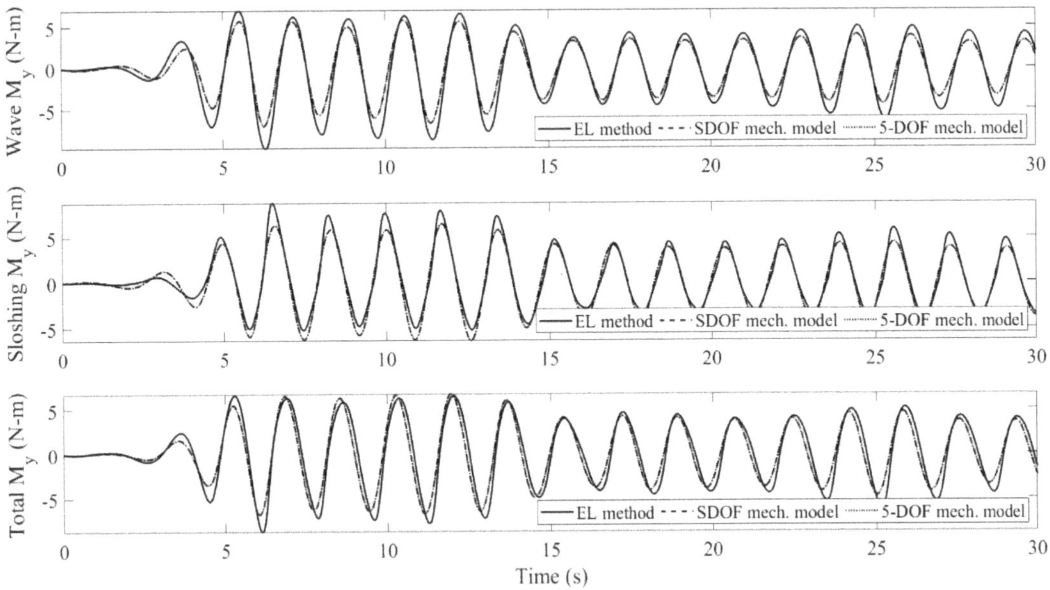

Figure 13. The overturning moments acting on the floating body for the large-wave case.

When the small wave excitation was applied, the body motions and the resultant forces along three directions are shown in Figures 14–19. The overall trends were similar to what was observed for the large-wave cases. However, the smaller wave loadings caused a smaller body vibration, hence lowering the nonlinear effect in the sloshing reactions. The mechanical model captured the pitch motion more easily. Similarly, the resultant forces and body dynamics by the SDOF and five-DOF mechanical systems highly coincided.

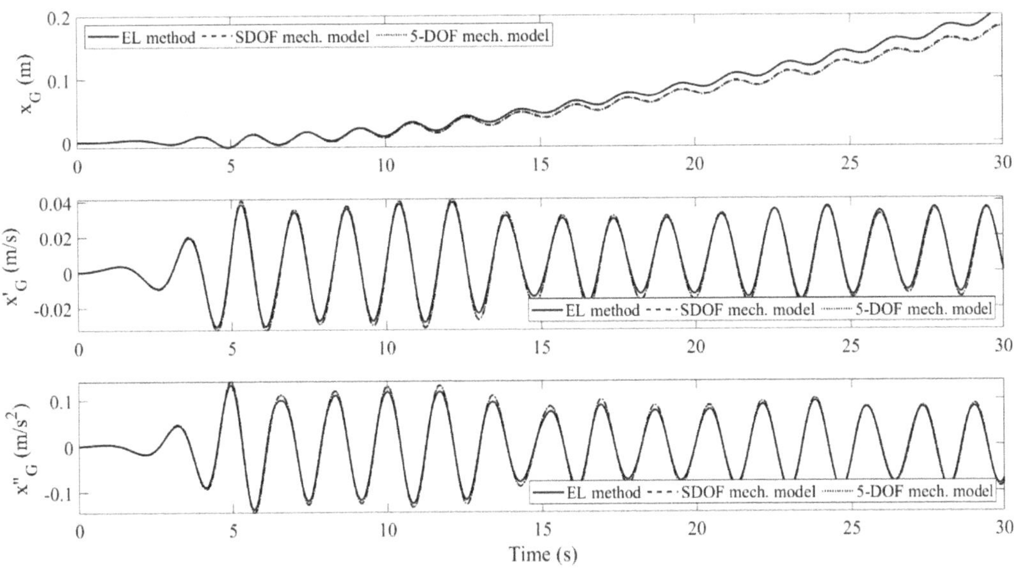

Figure 14. The surge motion responses of the floating body, configured with a water tank installed with porous media, undergoing small wave excitation.

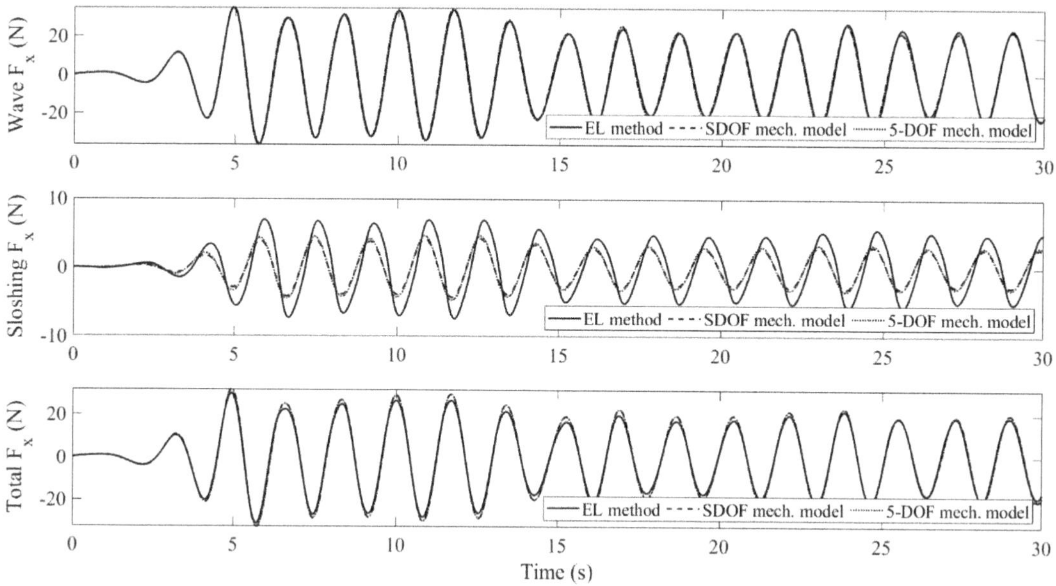

Figure 15. The horizontal forces acting on the floating body for the small-wave case.

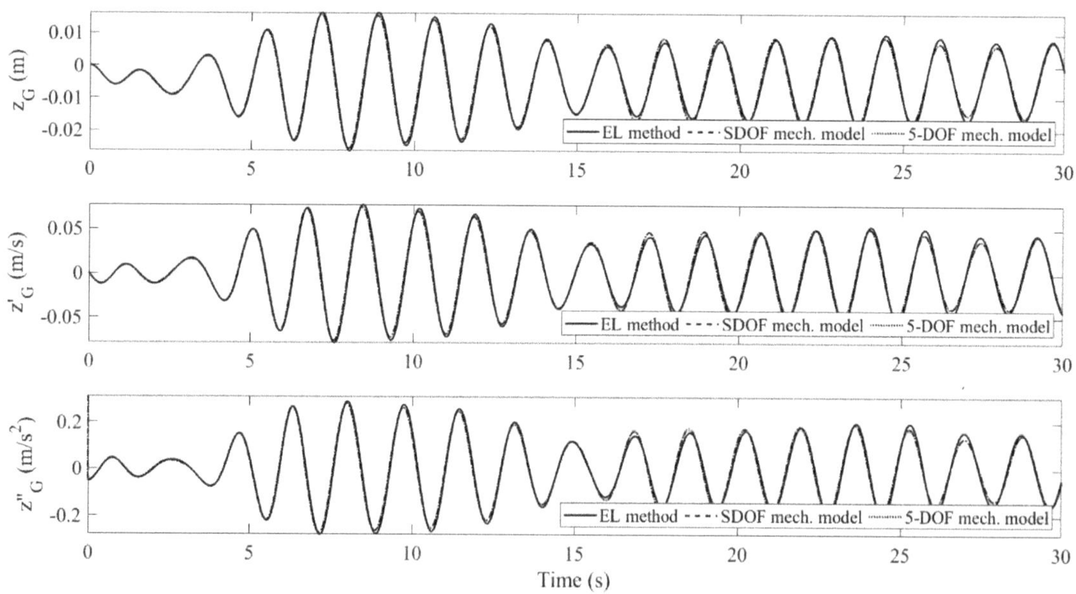

Figure 16. The heave motion responses of the floating body, configured with a water tank installed with porous media, undergoing small wave excitation.

The response spectra of the body motions for the large- and small-wave cases are shown in Figures 20 and 21, respectively. The spectra by both methods are very similar, hence the mechanical model ensured the consistency of the dynamic characteristics of the sloshing behavior. The response spectra by the SDOF and 5-DOF mechanical systems are also identical, so the importance of the first modal response was validated in the frequency domain. Both methods showed one peak at the resonant frequencies for each motion, but the EL method observed another local peak at the double frequency (1.16 Hz) of the angular response spectra. This means the nonlinear effect due to the convective term appeared. No peak at the double frequency was observed by the mechanical model because the nonlinear effect was neglected. However, this phenomenon was not obvious in the small-wave case as shown in Figure 21. Besides, the response spectra of the surge and heave motions had no local peak at the double frequency. The nonlinear sloshing response was negligible in these two vibration modes.

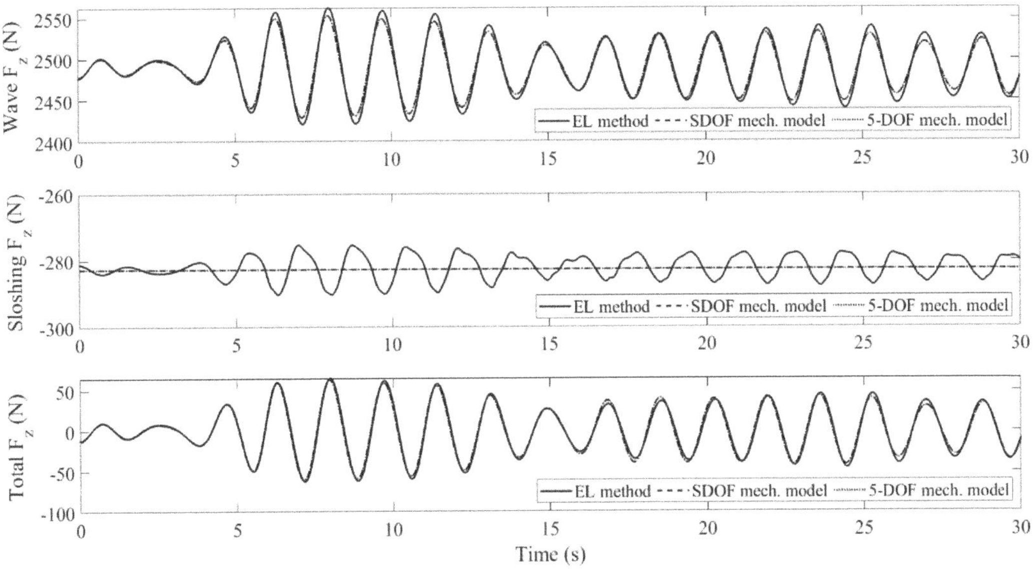

Figure 17. The vertical forces acting on the floating body for the small-wave case.

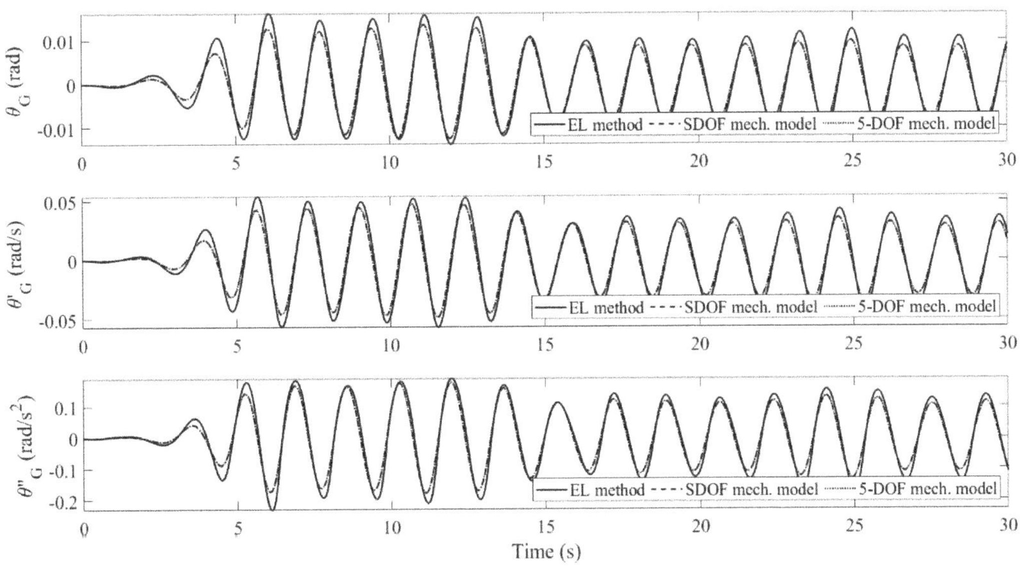

Figure 18. The pitch motion responses of the floating body, configured with a water tank installed with porous media, undergoing small wave excitation.

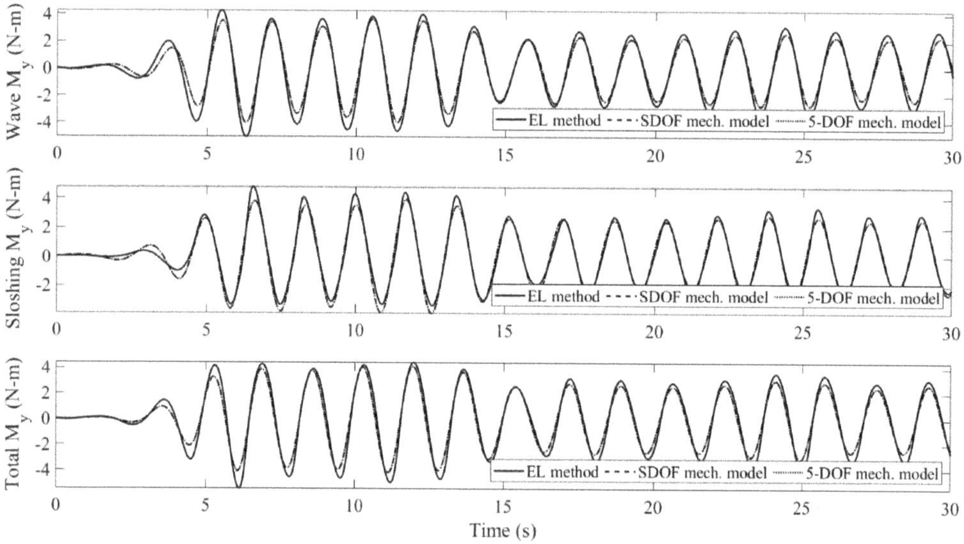

Figure 19. The overturning moments acting on the floating body for the small-wave case.

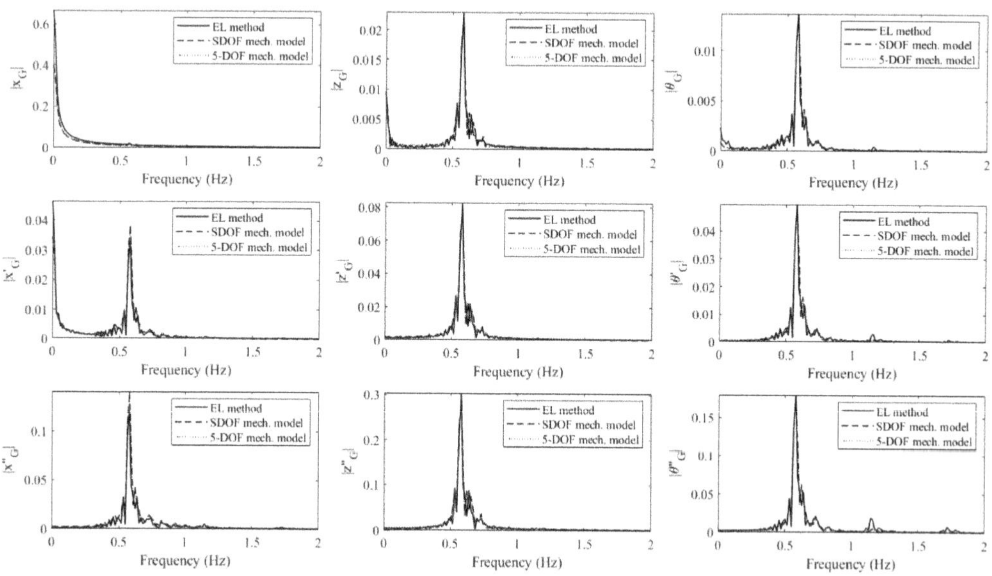

Figure 20. The response spectra of the floating body, configured with a water tank installed with porous media, undergoing large wave excitation.

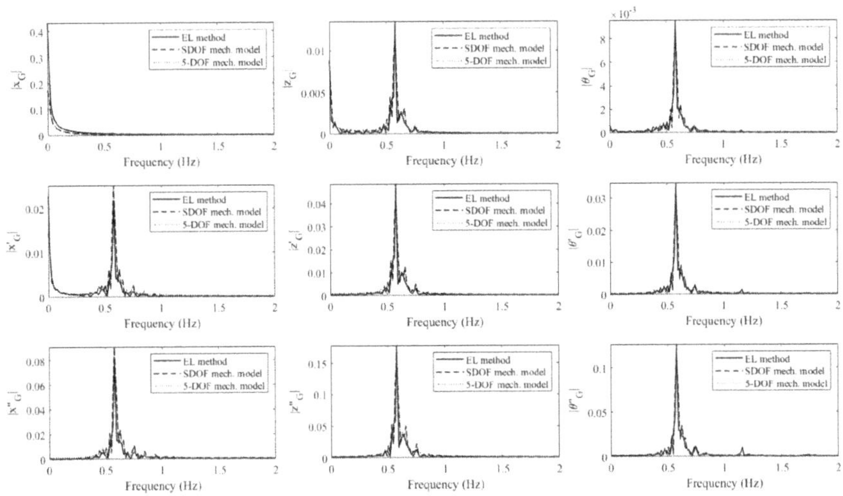

Figure 21. The response spectra of the floating body, configured with a water tank installed with porous media, undergoing small wave excitation.

The free-surface elevations on the left-side (stern side) wall of the tank for the large- and small-wave cases are shown in Figure 22. Note that in the mechanical model, the free-surface elevation of the sloshing fluid can be obtained by:

$$\eta = \sum_{n=0}^{\infty} \frac{4tanha_n h}{a_n L} sina_n x \cdot x_n \tag{51}$$

In the large-wave case, the EL method could simulate nonlinear waves. The mechanical model showed linear waves and overestimated the amplitude due to the lack of nonlinear damping. Similar results have been observed in past experiments [25]. In the small-wave case, the wave elevations by both methods were still different. However, it only slightly changed the sloshing forces. Besides, the five-DOF system included more high-mode sloshing responses than the SDOF system. It gave a higher hydrostatic pressure but those components were too small to affect the FSI simulation.

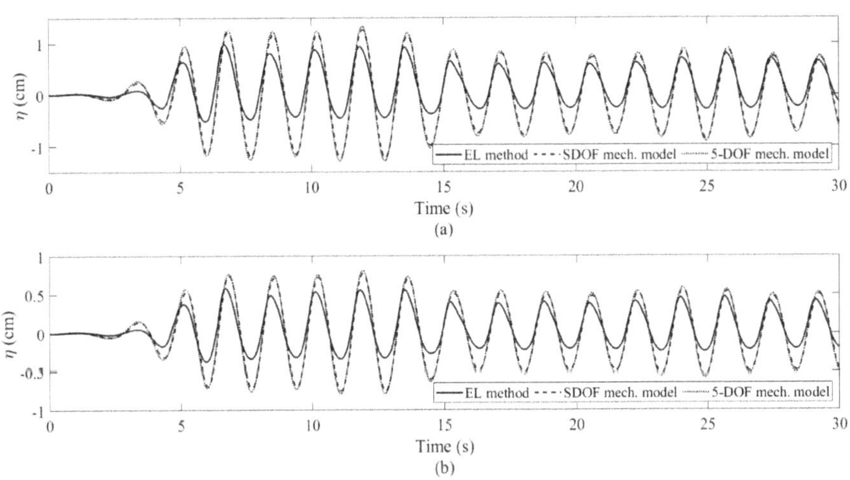

Figure 22. The free-surface elevations on the left-side wall of the tank for different wave loadings: (**a**) large wave; (**b**) small wave.

4.2. Floating Body in Resonant Waves: Controlled vs. Uncontrolled

The uncontrolled body means that no water tank was installed on the platform, otherwise, it is called a controlled body. For the controlled body, an additional case is considered where the tank was only filled with water and no porous media, which suggests the porosity $\gamma = 1$ and the coefficients $\alpha_1 = 0.1\ s^{-1}$ and $\alpha_2 = 0\ m^{-1}$ were used in the EL method [46]. When the large wave was applied, the body motions and the resultant forces along the horizontal direction are shown in Figures 23 and 24, respectively. The uncontrolled case stopped at 17.3 s due to the numerical instability induced by the high velocity of a wave jet on the body. The water tank was prone to keep the floating body oscillating around the original position, hence it could slow down the lateral drift. Although the force cancellation was weak, it still reduced the surge motion. However, the floating body was free to surge in our case, and the effectiveness of the surge mitigation by the porous media in the sloshing fluid was not obvious.

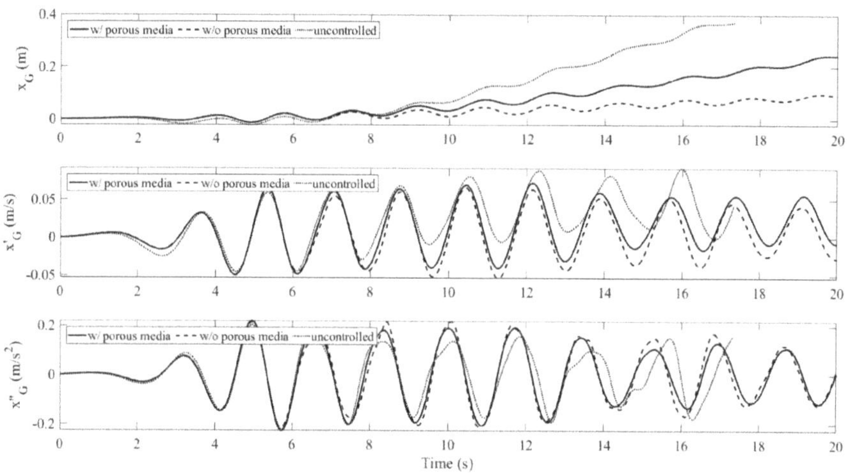

Figure 23. The surge responses of the controlled and uncontrolled floating bodies subject to large wave excitation.

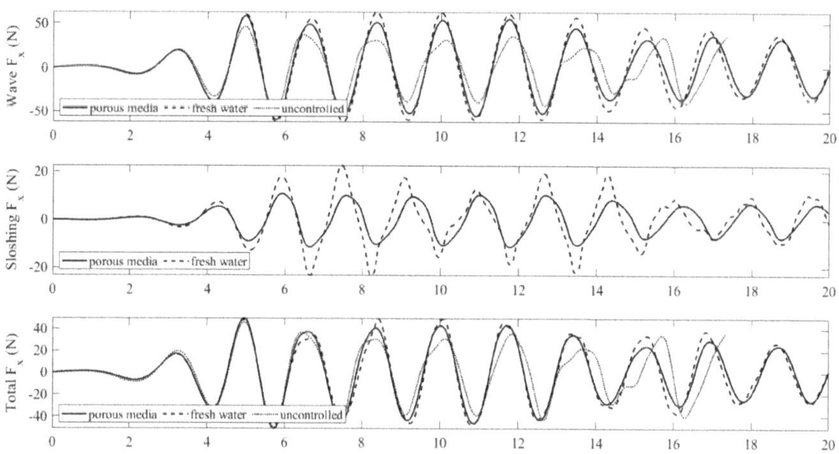

Figure 24. The horizontal forces act on the controlled and uncontrolled floating bodies for the large-wave case.

The body motions and the resultant forces along the vertical direction are shown in Figures 25 and 26, respectively. The mean values of the steady-state displacement were different for the controlled and uncontrolled cases because their total weights were different. The uncontrolled body was lighter, so it had a larger initial acceleration. As their transient response vanished, they oscillated in the same way but had different equilibrium points. This implies that the sloshing fluid did not change the heave motion of the body. Besides, the difference between the vertical sloshing forces from the tanks with and without porous media equaled their difference in water weight. The porous media did not affect the vertical sloshing force.

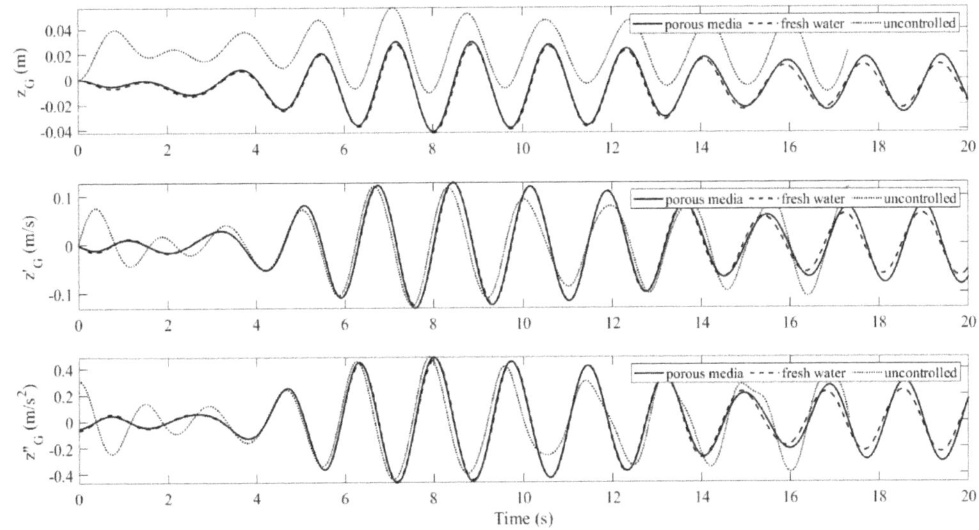

Figure 25. The heave responses of the controlled and uncontrolled floating bodies subject to large wave excitation.

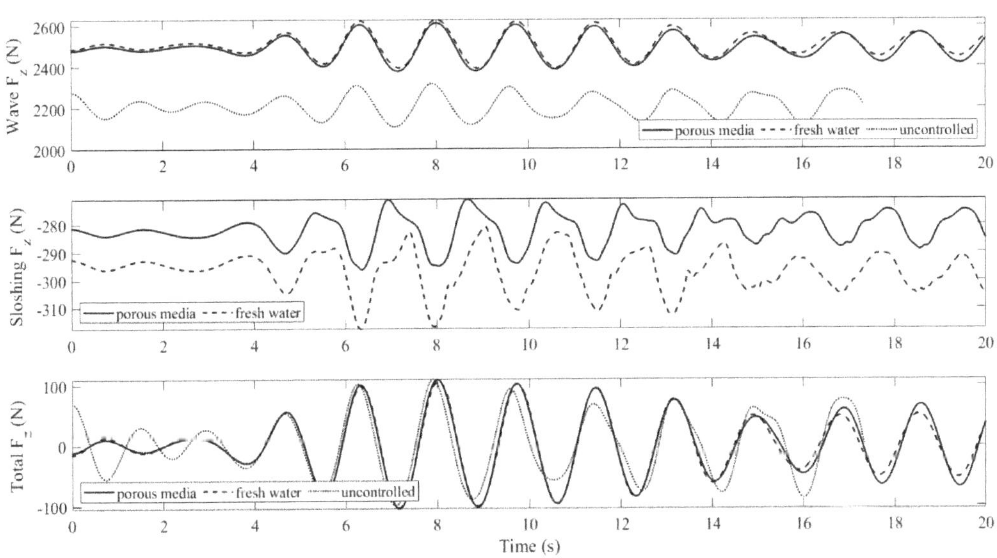

Figure 26. The vertical forces act on the controlled and uncontrolled floating bodies for the large-wave case.

The body motions and the resultant forces about the angular axis are shown in Figures 27 and 28, respectively. For the uncontrolled case, the responses were greatly amplified due to resonance. The maximum angular displacement was 11.3°. In the controlled cases, the maximum angular displacements were significantly reduced to 2.2° (without porous media) and 1.6° (with porous media). Moreover, their transient responses decayed very fast and the steady-state responses were suppressed. This can be explained later by the energy exchange in the body-tank system. The force cancellation between the sloshing reaction and wave loading was strong, showing that the tuning condition may have been satisfied. The tank with porous media had better performance in reducing pitch vibration than the other tank, but the overall performance should be validated through other near-resonance tests in the next example.

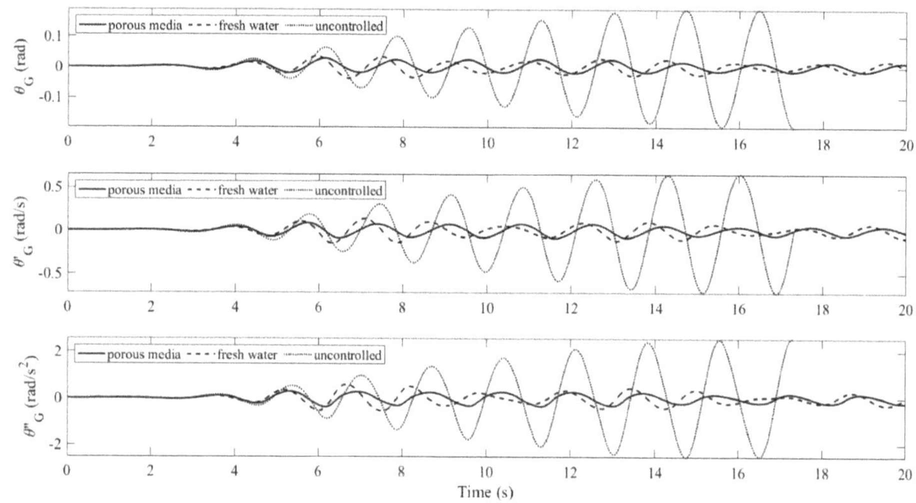

Figure 27. The pitch responses of the controlled and uncontrolled floating bodies subject to large wave excitation.

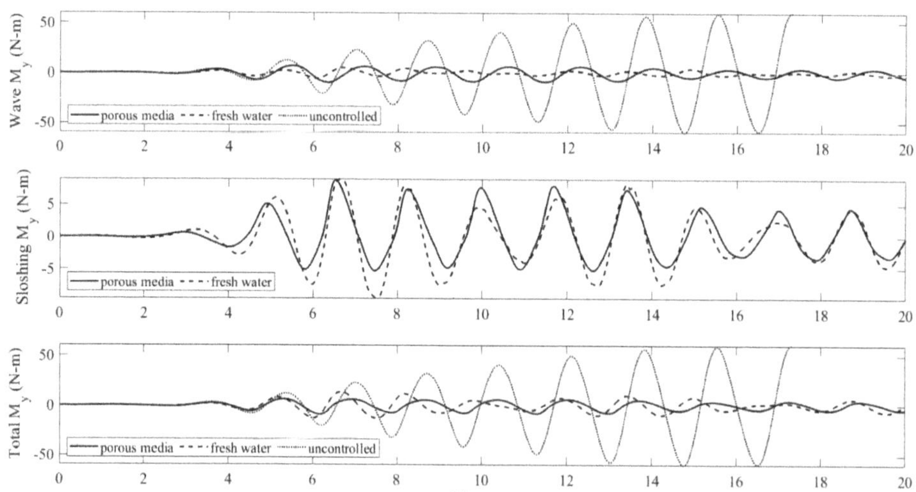

Figure 28. The overturning moments act on the controlled and uncontrolled floating bodies for the large-wave case.

The response spectra of the controlled and uncontrolled floating bodies subject to large wave excitation are shown in Figure 29. The major oscillatory frequency does not shift (still 0.58 Hz), while a small local peak shows up at the frequency of 0.53 Hz, which is equal to the natural frequency of the sloshing fluid in the tank. The dynamic characteristic of the entire system was not affected because the water tank was not heavy enough (the mass ratio was 13%). The maximum amplitudes of the angular responses of the floating body were greatly decreased. Compared to the tank without porous media, the porous media could further reduce the peak response of the surge and pitch motions.

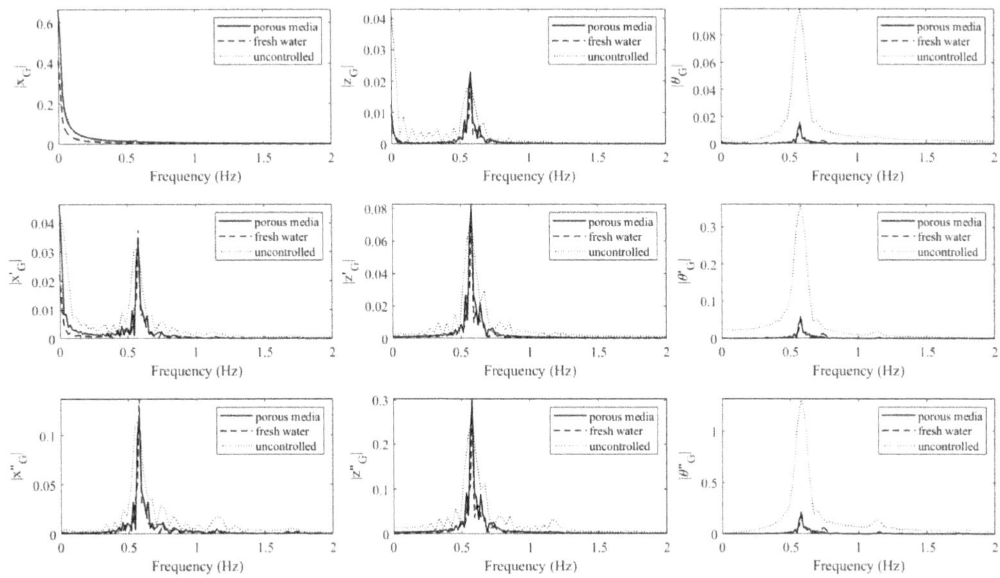

Figure 29. The response spectra of the controlled and uncontrolled floating bodies for the large-wave case.

The kinetic energy and potential energy of the floating body and sloshing fluid can be obtained by:

$$T_b = \frac{1}{2}M_b\left|\dot{x}_G\right|^2 + \frac{1}{2}I_b\left|\dot{\theta}_G\right|^2 \; and \; T_s = \frac{1}{2}\int_{S_1} \gamma\rho\phi\frac{\partial\phi}{\partial n}dS \tag{52}$$

$$V_b = M_b g z_G \; and \; V_s = \frac{1}{2}\gamma\rho g\int_{S_{f1}} \eta^2 dS \tag{53}$$

where the subscripts b and s denote the floating body and sloshing fluid, respectively. Note that T_s is obtained by applying Green's identity to the volume integral of the kinetic energy of a fluid particle [47]. The zero-potential plane is defined at $z = 0$. The changes in the kinetic and potential energy, i.e., the difference between the energy at current and initial states of the floating body are shown in Figure 30. In Figure 30a, the kinetic energy increased rapidly because the body velocity increased when the incident wave came. For the controlled case, the water tank dampened the transient vibrations, so the body's kinetic energy reached the steady state quickly. It is noticeable that the kinetic energy of the body without porous media was lower than that with porous media because of the lower heaving velocity, as shown in Figure 25. However, the porous media dampened more body kinetic energy from the surge and pitch components. Figures 25 and 30b show that the potential energy change was in sync with the vertical displacement. Since the water tank did not mitigate the heaving motion, the potential energy change of the body was irrelevant to the sloshing behavior and only depended on the vertical force equilibrium.

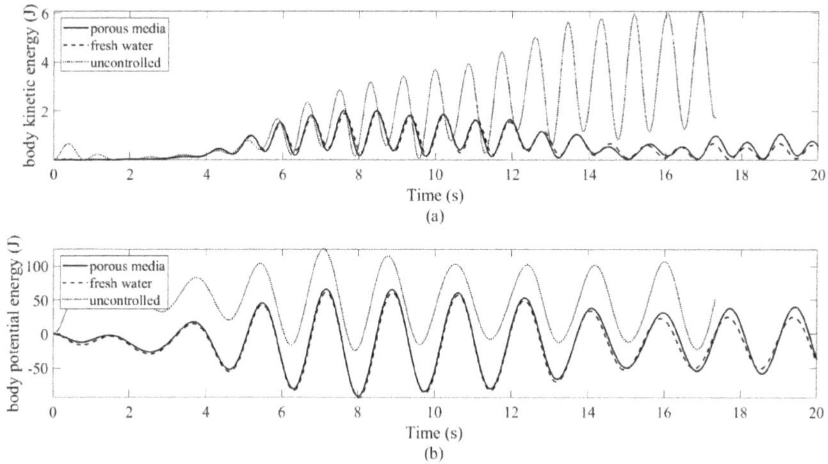

Figure 30. The energy changes of the floating body: (**a**) kinetic energy; (**b**) potential energy.

The kinetic and potential energy changes of the sloshing fluid are shown in Figure 31. For the first 15 s in Figure 31a, the kinetic energy of the sloshing fluid without porous media increased faster than that with porous media. This shows the porous media can quickly dampen the kinetic energy in the transient state, therefore suppressing the fluid velocity and the free-surface deformation. By correlating the force/moment and energy change as shown in Figures 24, 28, 30 and 31, the effect of porous media on the FSI can be characterized. When the incident waves hit the floating body, the work is transferred to the body's kinetic energy. Through the force cancellation, part of the wave-induced work will be taken by the sloshing fluid, and then dissipated by the damping mechanism of the porous media. Therefore, good force cancellation and enough sloshing damping are two key factors for structural vibration control. In Figure 31b, the potential energy of the fluid in the tank without porous media grew higher than that of the fluid in the porous media. Since they had similar elevations, the difference in the potential energy came from different free-surface deformations.

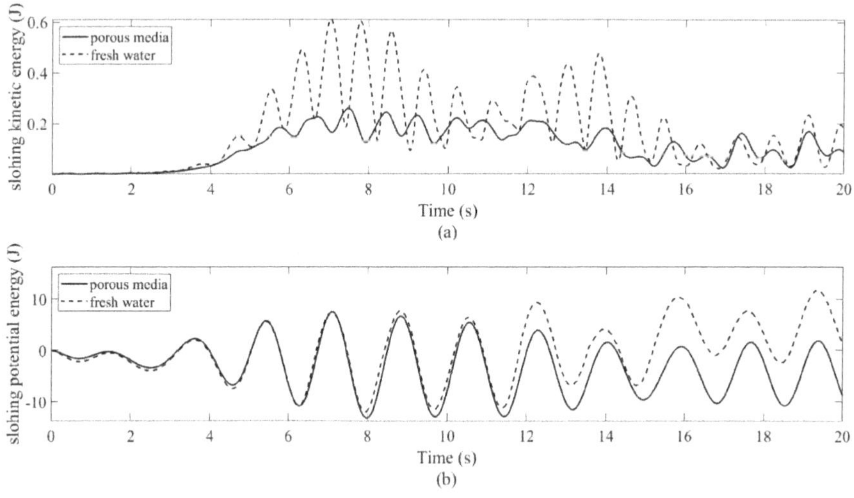

Figure 31. The energy changes of the sloshing fluid in the water tank: (**a**) kinetic energy; (**b**) potential energy.

The free-surface elevations on the left-side wall of the tank with and without porous media are shown in Figure 32. The wave elevation in the tank without porous media was very nonlinear, while the porous media reduced the degree of nonlinearity due to their extra damping effect. This shows the benefit for the liquid storage tank that not only the dynamic pressure can be reduced, but the roof damage can be alleviated.

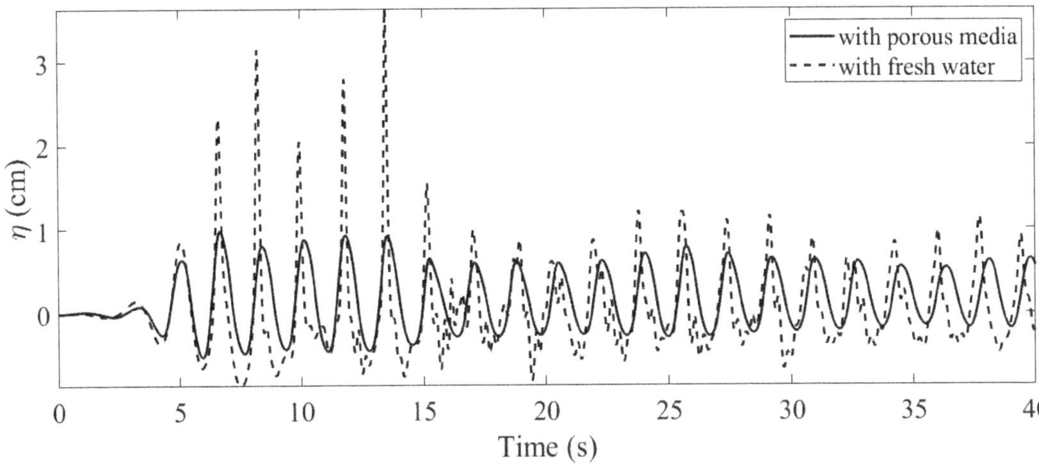

Figure 32. The wave elevations on the left-side wall of the tank with and without porous media.

The wave elevations of four gauges distributed at x = 2, 4, 6, and 8 m (two on the left side and two on the right side of the body) along the channel are shown in Figure 33. At the beginning ($t < 2$ s), the incident wave had not reached the body, so the body only oscillated vertically. The uncontrolled body had a larger initial acceleration, so it induced a larger disturbance to the waves, as shown in Figure 33b,c. The heave-induced wave that propagated downstream was recorded as is shown in Figure 33d. However, there was no significant change in the first gauge because of the overwhelming incident waves. As the incident wave reached the floating body ($t > 2$ s), the body started to surge and pitch. In our case, the wavelength was about 4.6 m, which was close to the distance between the wavemaker and the body. Therefore, a standing wave occurred so larger wave elevations at gauge two during t = 2~10 s were observed, as shown in Figure 33b. The influence on the far-distance wave field due to the body motion was negligible, as shown in Figure 33d.

4.3. Frequency Response Analysis

To complete the frequency response curves, the near-resonance condition was emphasized in the range of β = 0.85~1.15, where β is the ratio between the frequencies of the wavemaker's movement and the body's pitch motion. The small wave excitation was applied to avoid unstable numerical results in uncontrolled cases. For the first application of structural control, the water tank was installed as a TLD. The frequency responses of the root-mean-square (RMS) of the steady-state angular displacement and acceleration of the body are shown in Figure 34. TLDs can enhance the workability and comfort of the workstation. The average reduction in the RMS displacement and acceleration by the porous-media TLD was 85%, while that by the fresh-water TLD (TLD without porous media) was 80%. However, the porous-media TLD showed at least 10% better performance than the fresh-water one in the high-frequency range ($\beta \geq 1.075$). This is because the high nonlinearity of the fresh-water TLD can cause a strong mistuning effect. Once the natural frequency of the water tank deviates from the tuning frequency too much, the force cancellation will be weakened. In terms of the tuning condition, the porous-media TLD can be optimized more easily according to a simple design plan [22]. A single peak occurred at β = 1 in uncontrolled cases, while there were two peaks at β = 0.925 and 1.1

when the TLD was installed. Theoretically, the sloshing fluid had infinite degrees of freedom. Therefore, the body-tank system was a multi-degree-of-freedom system. Since the fundamental sloshing mode dominated (as was concluded in the previous example), the frequency response curves only had two peaks that represented two major vibration modes of the TLD-body system.

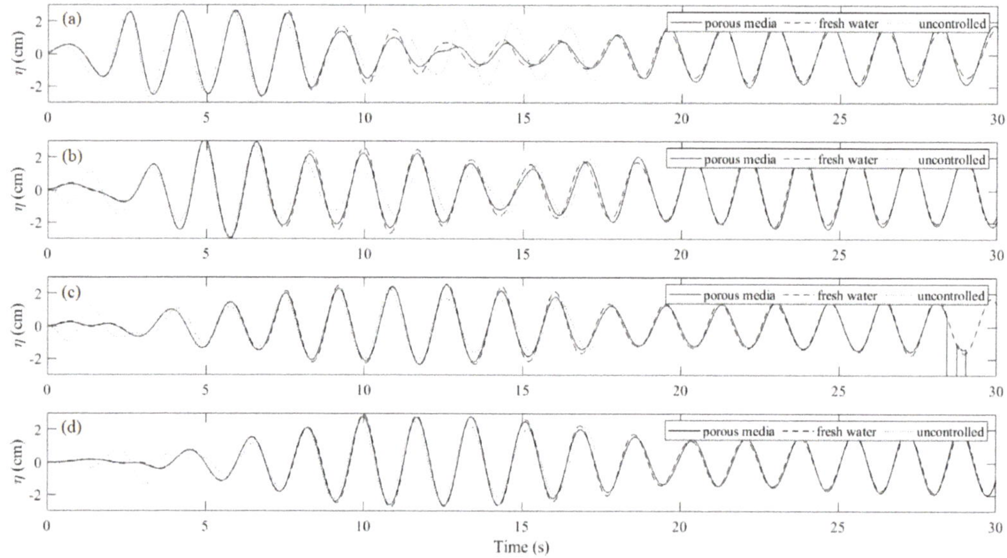

Figure 33. The wave elevations in the channel at the locations of: (**a**) $x = 2$ m; (**b**) $x = 4$ m; (**c**) $x = 6$ m; (**d**) $x = 8$ m.

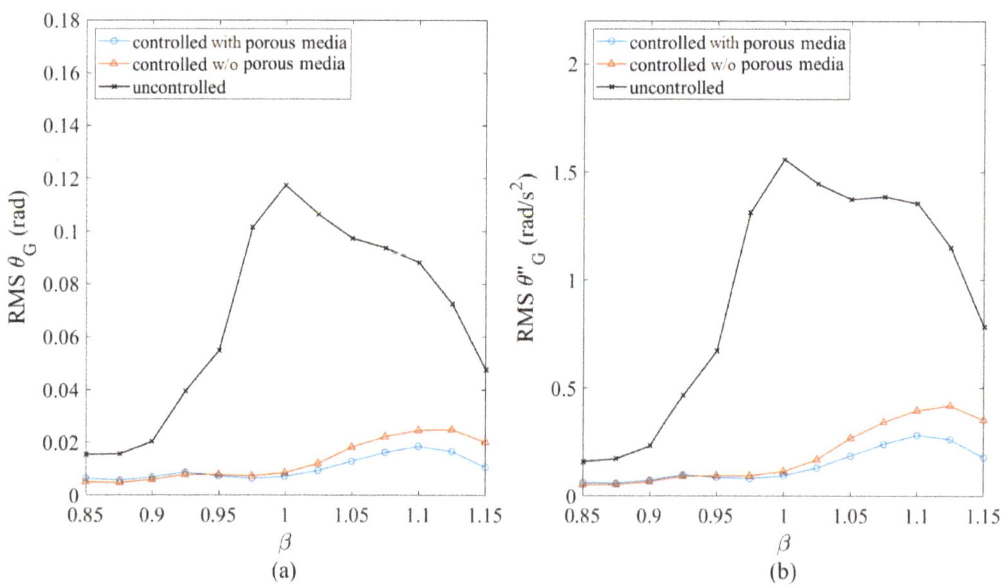

Figure 34. The frequency responses of the RMS responses of the controlled and uncontrolled floating bodies for the small-wave case: (**a**) angular displacement; (**b**) angular acceleration.

For a liquid carrier on the ocean, its stability is related to the amplitude of the wave-induced sloshing resultants. The frequency responses of the maximum horizontal force and overturning moment due to sloshing fluid are shown in Figure 35. In Figure 35a, the porous media can reduce the horizontal force by 30~70% because the extra damping effect limits the sloshing responses, hence reducing the hydrodynamic force. In Figure 35b, the reduction in the overturning moment was low. This is because, for a shallow-water tank used in this example, the overturning moment did not mainly come from the horizontal component. Therefore, the reduction by the porous media was not obvious. The integrity of a liquid container is another important issue that is related to pressure concentration. The frequency responses of the maximum dynamic pressure on the wetted wall and the normalized height of the local maximum pressure are shown in Figure 36. Note that h_p denotes the elevation of the fluid particle that has the maximum dynamic pressure. In Figure 36a, the maximum dynamic pressure in the tank without porous media was 485 Pa, which was higher than the maximum hydrostatic pressure (460 Pa). The porous media reduced 25~45% of the dynamic pressure. The averages were 343 versus 228 Pa. In Figure 36b, the elevations of the local maximum pressure were similar, which were below the still-water line ($h_p/h < 1$) in most cases. As the wave traveled back and forth during the vibration, the greatest impact occurred at the instant when it reached the walls. The contact location was usually near but below the still-water line. This phenomenon was visualized in the past article [48]. Overall, the porous media will not change the location of the pressure concentration but can effectively reduce the hydrodynamic pressure.

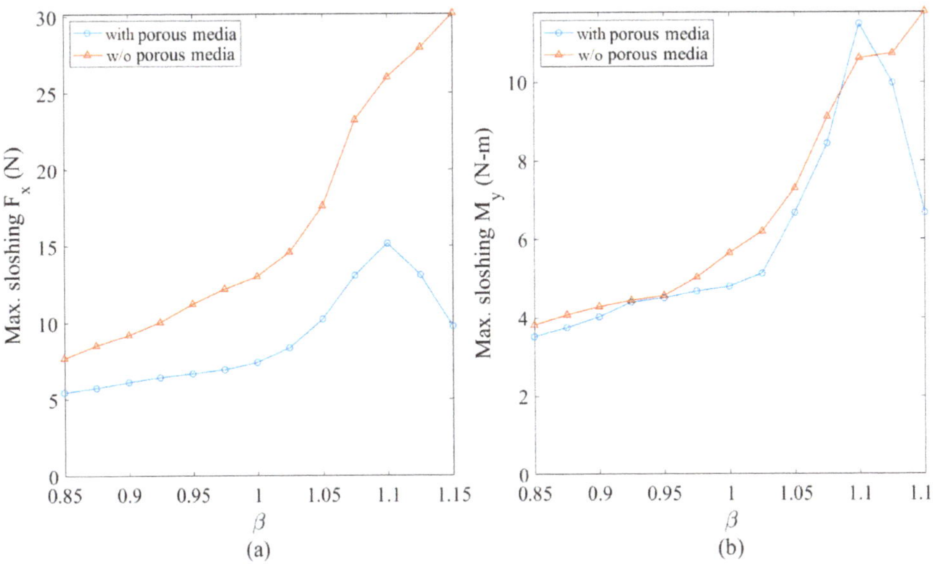

Figure 35. The frequency responses of the maximum sloshing reactions for the small-wave case: (**a**) horizontal force; (**b**) overturning moment.

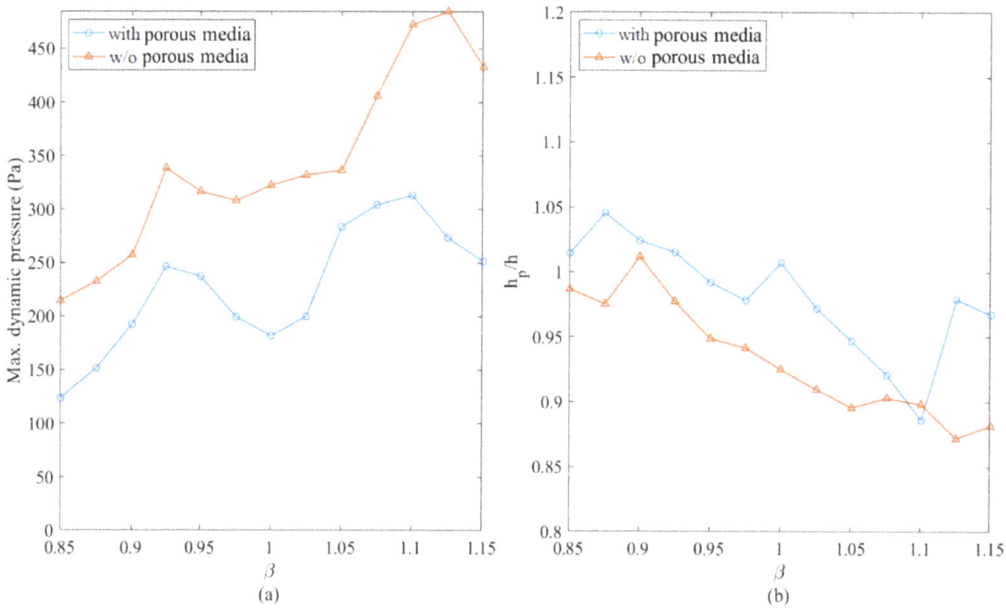

Figure 36. The frequency responses of the local sloshing behavior in the water tank on a floating body for the small-wave case: (**a**) maximum dynamic pressure on the wetted wall; (**b**) normalized height of pressure concentration.

5. Conclusions

From this study, some conclusions can be drawn:

1. Porous media can accelerate the energy transmission in the FSI process, enhance the damping capability of the sloshing fluid, and reduce the sloshing velocity and deformation.
2. The sloshing fluid will create a control force on the body. Under certain circumstances, it will cancel the wave loading. When it happens, the work done by incident waves is transferred to the fluid in the water tank (meanwhile, the wave-induced sloshing behavior appears), and then dissipated by the damping mechanism of porous media.
3. The mechanical model produces linear sloshing reactions, which affects the accuracy of the amplitude of the body's pitch motion for the large-wave case. However, it still captures the oscillatory frequencies very well. For the small-wave case, the difference is insignificant. When the tank is subject to pitch motion, the first modal sloshing response dominates.
4. In the near-resonance tests, the average reduction in the RMS pitch motion by the porous-media TLD was 85%. It did not produce an effective control force for the heave motion. Moreover, the force cancellation in the horizontal motion was weak. Therefore, the application in surge mitigation is uncertain. Compared to the fresh-water TLD, the mistuning effect was alleviated.
5. Porous media can effectively decrease the hydrodynamic pressure by up to 45%. The integrity of the liquid container and the stability of the liquid carrier can be improved.

Author Contributions: Conceptualization, W.-H.T. and C.E.K.; methodology, W.-H.T. and C.E.K.; software, W.-H.T.; formal analysis, W.-H.T.; investigation, W.-H.T. and Y.-C.C.; data curation, W.-H.T.; writing—original draft preparation, W.-H.T. and Y.-C.C.; writing—review and editing, W.-H.T., Y.-C.C., C.E.K. and L.M.; visualization, W.-H.T.; supervision, C.E.K. and L.M.; funding acquisition, C.E.K. All authors have read and agreed to the published version of the manuscript.

Funding: This work was partially supported by internal research funding from the Center for Computation and Technology and the College of Engineering at Louisiana State University.

Institutional Review Board Statement: Not applicable.

Informed Consent Statement: Not applicable.

Data Availability Statement: All data can be reproduced by the Fortran code in the public repository https://github.com/whtsao/PMTLD-in-waves, accessed on 1 May 2022.

Acknowledgments: Portions of this research were conducted with high-performance computing resources provided by Louisiana State University (http://www.hpc.lsu.edu, accessed on 10 January 2021).

Conflicts of Interest: The authors declare no conflict of interest.

References

1. Cariou, A.; Casella, G. Liquid sloshing in ship tanks: A comparative study of numerical simulation. *Mar. Struct.* **1999**, *12*, 183–198. [CrossRef]
2. Kim, Y.; Nam, B.W.; Kim, D.W.; Kim, Y.S. Study on coupling effects of ship motion and sloshing. *Ocean Eng.* **2007**, *34*, 2176–2187. [CrossRef]
3. Lee, S.J.; Kim, M.H.; Lee, D.H.; Kim, J.W.; Kim, Y.H. The effects of LNG-tank sloshing on the global motions of LNG carriers. *Ocean Eng.* **2007**, *34*, 10–20. [CrossRef]
4. Lyu, W.J.; Riesner, M.; Peters, A.; Moctar, O. A hybrid method for ship response coupled with sloshing in partially filled tanks. *Mar. Struct.* **2019**, *67*, 102643. [CrossRef]
5. Burger, W.; Corbet, A.G. Chapter III—Anti-Rolling Devices in General. In *Ship Stabilizers: A Handbook for Merchant Navy Officers*; 1966; pp. 31–37. Available online: https://www.sciencedirect.com/science/article/pii/B9780080115047500095 (accessed on 1 May 2022).
6. Chen, Y.H.; Ko, C.H. Active tuned liquid column damper with propellers. *Earthq. Eng. Struct. Dyn.* **2003**, *32*, 1627–1638. [CrossRef]
7. Yalla, S.K.; Kareem, A.; Kantor, J.C. Semi-active tuned liquid column dampers for vibration control of structures. *Eng. Struct.* **2001**, *23*, 1469–1479. [CrossRef]
8. Marzouk, O.A.; Nayfeh, A.H. Control of ship roll using passive and active anti-roll tanks. *Ocean Eng.* **2009**, *36*, 661–671. [CrossRef]
9. Gattulli, V.; Ghanem, R. Adaptive control of flow-induced oscillations including vortex effects. *Int. J. Non-Linear Mech.* **1999**, *34*, 853–868. [CrossRef]
10. Kawano, K. Active control effects on dynamic response of offshore structures. In Proceedings of the Third International Offshore and Polar Engineering Conference, Singapore, 6–11 June 1993; pp. 594–598.
11. Stewart, G.; Lackner, M. Offshore wind turbine load reduction employing optimal passive tuned mass damping systems. *IEEE Trans. Control Syst. Technol.* **2013**, *21*, 1090–1104. [CrossRef]
12. Si, Y.; Karimi, H.R.; Gao, H. Modelling and optimization of a passive structural control design for a spar-type floating wind turbine. *Eng. Struct.* **2014**, *69*, 168–182. [CrossRef]
13. Dinh, V.N.; Basu, B. Passive control of floating offshore wind turbine nacelle and spar vibrations by multiple tuned mass dampers. *Struct. Control Health Monit.* **2014**, *22*, 152–176. [CrossRef]
14. Tong, X.; Zhao, X.; Karcanias, A. Passive vibration control of an offshore floating hydrostatic wind turbine model. *Wind Energy* **2018**, *21*, 697–714. [CrossRef]
15. Chen, J.; Zhan, G.; Zhao, Y. Application of spherical tuned liquid damper in vibration control of wind turbine due to earthquake excitations. *Struct. Des. Tall Spec. Build.* **2016**, *25*, 431–443. [CrossRef]
16. Chen, Y.H.; Hwang, W.S.; Tsao, W.H. Nonlinear dynamic characteristics of rectangular and cylindrical TLD's. *J. Eng. Mech.* **2018**, *144*, 06018004. [CrossRef]
17. Zahrai, S.M.; Abbasi, S.; Samali, B.; Vrcelj, Z. Experimental investigation of utilizing TLD with baffles in a scaled down 5-story benchmark building. *J. Fluids Struct.* **2012**, *28*, 194–210. [CrossRef]
18. Warnitchai, P.; Pinkaew, T. Modelling of liquid sloshing in rectangular tanks with flow-dampening devices. *Eng. Struct.* **1998**, *20*, 293–600. [CrossRef]
19. Faltinsen, O.M.; Firoozkoohi, R.; Timokha, A.N. Analytical modeling of liquid sloshing in a two-dimensional rectangular tank with a slat screen. *J. Eng. Math.* **2011**, *70*, 93–109. [CrossRef]
20. Hamelin, J.; Love, J.; Tait, M.; Wison, J. Tuned liquid dampers with a Keulegan-Carpenter number-dependent screen drag coefficient. *J. Fluids Struct.* **2013**, *43*, 271–286. [CrossRef]
21. Tait, M.J. Modelling and preliminary design of a structure-TLD system. *Eng. Struct.* **2008**, *30*, 2644–2655. [CrossRef]
22. Tsao, W.H.; Hwang, W.S. Tuned liquid dampers with porous media. *Ocean Eng.* **2018**, *167*, 55–64. [CrossRef]
23. Tsao, W.H.; Hwang, W.S. Dynamic characteristics of liquid sloshing in cylindrical tanks filled with porous media. In *IOP Conference Series: Earth and Environmental Science*; IOP Publishing: Bristol, UK, 2019; Volume 351.
24. Tsao, W.H.; Chang, T.J. Sloshing phenomenon in rectangular and cylindrical tanks filled with porous media: Supplementary solution and impulsive-excitation experiment. *J. Eng. Mech.* **2020**, *146*, 04020139. [CrossRef]
25. Tsao, W.H.; Huang, L.H.; Hwang, W.S. An equivalent mechanical model with nonlinear damping for sloshing rectangular tank with porous media. *Ocean Eng.* **2021**, *242*, 110145. [CrossRef]

26. Tsao, W.H.; Huang, Y.L. Sloshing force in a rectangular tank with porous media. *Results Eng.* **2021**, *11*, 100250. [CrossRef]
27. Sen, D. Numerical simulation of motions of two-dimensional floating bodies. *J. Ship Res.* **1993**, *37*, 307–330. [CrossRef]
28. Van Daalen, E.F.G. Numerical and Theoretical Studies of Water Waves and Floating Bodies. Ph.D. Thesis, Universiteit Twente, Enschede, The Netherlands, 1993.
29. Tanizawa, K. A nonlinear simulation method of 3D body motions in waves. *J. Soc. Nav. Archit. Jpn.* **1995**, *178*, 179–191. [CrossRef]
30. Jung, K.H.; Chang, K.A.; Jo, H.J. Viscous effect on the roll motion of a rectangular structure. *J. Eng. Mech.* **2006**, *132*, 190–200. [CrossRef]
31. Guerber, E.; Benoit, M.; Grilli, S.T.; Buvat, C. A fully nonlinear implicit model for wave interactions with submerged structures in forced or free motion. *Eng. Anal. Bound. Elem.* **2012**, *36*, 1151–1163. [CrossRef]
32. Dombre, E.; Benoit, M.; Violeau, D.; Peyrard, C.; Grilli, S.T. Simulation of floating structure dynamics in waves by implicit coupling of a fully non-linear potential flow model and a rigid body motion approach. *J. Ocean Eng. Mar. Energy* **2015**, *1*, 55–76. [CrossRef]
33. Vinje, T.; Brevig, P. *Nonlinear, Two-Dimensional Ship Motions*; Technical Report; Norwegian Institute of Technology: Trondheim, Norway, 1981; p. R-112.81.
34. Belibassakis, K.A. A boundary element method for the hydrodynamic analysis of floating bodies in variable bathymetry regions. *Eng. Anal. Bound. Elem.* **2008**, *32*, 796–810. [CrossRef]
35. Wu, G.X.; Eatock-Taylor, R. Transient motion of a floating body in steep water waves. In Proceedings of the 11th Workshop on Water Waves and Floating Bodies, Hamburg, Germany, 17–20 March 1996.
36. Matthies, H.G.; Niekamp, R.; Steindorf, J. Algorithms for strong coupling procedures. *Comput. Methods Appl. Mech. Eng.* **2006**, *195*, 2028–2049. [CrossRef]
37. Palm, J.; Eskilsson, C.; Paredes, G.M.; Bergdahl, L. Coupled mooring analysis of floating wave energy converters using CFD: Formulation and Validation. *Int. J. Mar. Energy* **2016**, *16*, 83–99. [CrossRef]
38. Tezduyar, T.E. Finite element methods for flow problems with moving boundaries and interfaces. *Comput. Methods Eng.* **2001**, *8*, 83–130. [CrossRef]
39. Rakhsha, M.; Kees, C.E.; Negrut, D. Lagrangian vs. Eulerian: An analysis of two solution methods for free-surface flows and fluid solid interaction problems. *Fluids* **2021**, *6*, 460. [CrossRef]
40. Karimirad, M.; Michailides, C.; Nematbakhsh, A. Offshore Mechanics: Structural and Fluid Dynamics for Recent Applications. 2018. Available online: https://www.wiley.com/en-us/Offshore+Mechanics%3A+Structural+and+Fluid+Dynamics+for+Recent+Applications-p-9781119216629 (accessed on 1 May 2022).
41. Grilli, S.T.; Skourup, J.; Svendsen, I.A. An efficient boundary element method for nonlinear water waves. *Eng. Anal. Bound. Elem.* **1989**, *6*, 97–107. [CrossRef]
42. Grilli, S.T.; Horrillo, J. Numerical generation and absorption of fully nonlinear periodic waves. *J. Eng. Mech.* **1997**, *123*, 1060–1069. [CrossRef]
43. Tsao, W.H.; Kinnas, S.A. Local simulation of sloshing jet in a rolling tank by viscous-inviscid interaction method. *Results Eng.* **2021**, *11*, 100270. [CrossRef]
44. Nield, A.D.; Bejan, A. *Convection in Porous Media*, 4th ed.; Springer: New York, NY, USA, 2013.
45. Ursell, F.; Dean, R.G.; Yu, Y.S. Forced small-amplitude water waves: A comparison of theory and experiment. *J. Fluid Mech.* **1960**, *7*, 33–52. [CrossRef]
46. Chen, Y.H.; Hwang, W.S.; Tsao, W.H. Nonlinear sloshing analysis by regularized boundary integral method. *J. Eng. Mech.* **2017**, *143*, 040170046. [CrossRef]
47. Currie, I.G. *Fundamental Mechanics of Fluids*, 3rd ed.; CRC Press: Boca Raton, FL, USA, 2003.
48. Delorme, L.; Colagrossi, A.; Souto-Iglesias, A.; Zamora-Rodríguez, R.; Botia-Vera, E. A set of canonical problems in sloshing, Part I: Pressure field in forced roll comparison between experimental results and SPH. *Ocean Eng.* **2009**, *36*, 168–178. [CrossRef]

applied sciences

MDPI

Article

Machine Learning Techniques Applied to Identify the Two-Phase Flow Pattern in Porous Media Based on Signal Analysis

Xiangyu Li, Liangxing Li *, Wenjie Wang, Haoxiang Zhao and Jiayuan Zhao

State Key Laboratory of Multiphase Flow in Power Engineering, Xi'an Jiaotong University, Xian-Ning West Road 28#, Xi'an 710049, China
* Correspondence: liangxing.li@xjtu.edu.cn; Tel.: +86-15129088385

Abstract: The development of flow pattern identification technology for gas–liquid two-phase flow in porous media is of great significance to engineering research and production. In this paper, a high accuracy identification method for two-phase flow pattern in porous media is proposed with Machine learning techniques. The gas–liquid two-phase flow patterns and corresponding differential pressure signals in porous beds with particle diameters of 1.5 mm, 3 mm, and 6 mm are obtained through visual experiments. Three time domain characteristic parameters (Mean, Standard deviation, and Range) are calculated by a statistical method, while the EMD energy spectrum of the signal is obtained by empirical mode decomposition. Based on these parameters, machine learning technology, including support vector machine (SVM) and BP neural network, are employed to identify the flow pattern. Four flow pattern identification models are trained based on SVM and BP neural network, with accuracies of 94.77%, 93.4%, 96.08%, and 91.5%. Furthermore, the three models with good performance are integrated by integrated learning technology. An integrated identification model of gas–liquid two-phase flow pattern in porous media with an overall accuracy of 98.04% is finally obtained.

Keywords: flow patterns identification; porous media; two-phase flow; support vector machine; BP network; feature extraction

Citation: Li, X.; Li, L.; Wang, W.; Zhao, H.; Zhao, J. Machine Learning Techniques Applied to Identify the Two-Phase Flow Pattern in Porous Media Based on Signal Analysis. *Appl. Sci.* **2022**, *12*, 8575. https://doi.org/10.3390/app12178575

Academic Editor: Bernhard Schrefler

Received: 29 July 2022
Accepted: 24 August 2022
Published: 27 August 2022

1. Introduction

Two-phase flow in porous media plays a critical role in scientific and industrial processes. During a severe accident of a light water reactor (LWR), the corium may form a debris bed with a porous media structure [1]. Understanding the two-phase flow mechanism in porous media is important to developing efficient cooling technology and terminating severe accidents. In petroleum engineering applications, since the oil–water mixtures are stored in the underground cracks of porous media structures, the study of two-phase flow in porous media is conducive to the development of more efficient oil extraction technologies [2]. In addition, in agriculture engineering [3], chemical engineering [4], and other fields, two-phase flow in porous media also widely exists and has an important impact on experiments and production. Flow pattern, which has a great effect on heat and mass transfer, is the research foundation to clarify the flow and heat transfer mechanism of two-phase flow in porous media as well as improve industrial production efficiency [5]. Accurate identification of flow patterns is the premise of flow pattern research. For the needs of experiments and industrial production, it is of great significance to develop accurate and fast flow pattern identification technology for porous media.

In early research, scholars directly observe the two-phase flow patterns in transparent pipes (including pipes filled with porous media) with their eyes. They distinguish different flow patterns according to the forms of the gas–liquid phase [6,7]. Afterward, the application of high-speed cameras helps the researchers capture the subtle changes in flow patterns timely and accurately [8]. This direct observation method is very effective for classifying and identifying typical flow patterns. However, it also has some limitations. For example, in real production, there are few transparent pipelines, which limits direct observation. In addition, for the transition flow pattern between typical flow patterns, it is difficult to distinguish which flow pattern it belongs to only through observation since it will be disturbed by the subjectivity of the observer.

In order to solve the above problems, scholars try to improve the flow pattern identification method. Some scholars try to establish experimental systems in the laboratory to simulate the real situation. Using these devices, scholars have observed various flow patterns and obtained some corresponding signals that are easy to be measured, such as differential pressure signals [9–11], temperature signals [12], and so on. After extracting the features of these signals and classifying them according to the corresponding flow patterns, the relationship between different flow patterns and signal features can be established, which is the basis for flow pattern judgment. The signal feature extraction techniques can be divided into three categories: time domain feature extraction, frequency domain feature extraction, and time-frequency domain feature extraction. Time domain feature extraction includes mean value, variance, and probability density function (PDF), etc. The common method of frequency domain feature extraction is power spectral density (PSD). Time-frequency analysis technology appears late, including wavelet analysis, empirical mode decomposition (EMD), and Hilbert-Huang transform [13]. Matsui [14] calculated the probability density function (PDF) of the differential pressure signal and distinguished the flow pattern by the shape of the PDF curve. Elperin and Klochko [15] extracted the wavelet energy spectrums of differential pressure signals to distinguish the flow patterns. dos Reis [16] summarized the characteristics of power spectral density (PSD) and PDF of the capacitive probe signal of slug flow. Li [17] identified the flow pattern in the dust collector through the synthesis of PSD, PDF, and wavelet features. Wu [10] characterized different flow regimes and their transition in a concentric vertical annulus through autocorrelation, PSD, Shannon entropy, and permutation entropy.

In recent years, machine learning technology has been developed in industries greatly [18–21]. It shows a strong ability in pattern recognition and classification, which provides a reference for the progress of flow pattern identification technology. More and more scholars have applied machine learning technology to the field of flow pattern recognition and developed a series of well-performing flow pattern recognition methods. Liang [22] used ultrasonic echoes and RBF neural network to identify the flow patterns in a horizontal pipe. Pei [23] utilized the complex network theory to identify the flow patterns in water pipelines. Guo [12] applied a neural network with the temperature fluctuation on pipe walls in classifying flow patterns. In addition to neural networks, support vector machine (SVM) technology is also widely used in flow pattern recognition. Zhang [24] identified the oil–gas two phase flow pattern based on SVM and electrical capacitance tomography technique. Liu [25] used doppler spectrum analysis and SVM to identify the flow pattern of oil–water two-phase flow. Ambrosio [26] used void fraction time series and SVM to classify the two-phase flow pattern in a vertical pipe.

It can be seen that the flow pattern recognition method combining signal feature extraction and machine learning technology has been introduced on some occasions. Unfortunately, compared with the importance of two-phase flow pattern identification in porous media, the research on this technology is far from enough. The present study proposes a new method to identify the gas–liquid two-phase flow pattern in porous media based on differential pressure signals and machine learning technology. The time domain characteristics and time-frequency domain (EMD) characteristics of the differential pressure signal are obtained. Using these features, a variety of flow pattern identification models based on SVM and BP neural networks are constructed. The comparison of different models verifies the performance of SVM and BP neural networks in porous media flow pattern recognition. Finally, an intelligent online flow pattern identification system for porous media is constructed by using three optimal models and integrated recognition technology.

Compared with the single model, the identification ability of the system has been improved. This paper provides a new idea for the development of flow pattern identification methods in porous media in the fields of the chemical industry, agriculture, petroleum, and nuclear engineering.

2. Description of Experiments

The experiments are carried out on a visual experimental system named DEBECO-LT (Debris Bed Coolability-Low Temperature). Figure 1 shows the diagram of the experimental system. In this paper, air and water are sent to the bottom of the test section, where they mix and flow into the porous media pipe. The wall of the experimental section is made of plexiglass, and two stainless steel wire meshes are fixed at the top and bottom to ensure that the glass balls will not move casually. Rosemont-3051 transmitters (0.04%) are fixed at the inlet and outlet of the test section to measure the differential pressure signals; the signal acquisition frequency is 400 Hz. FastcamMini Photron high-speed camera (102, 400 frame/s) is used to capture clear flow pattern photos. The previous paper describes the experimental system in more detail [27].

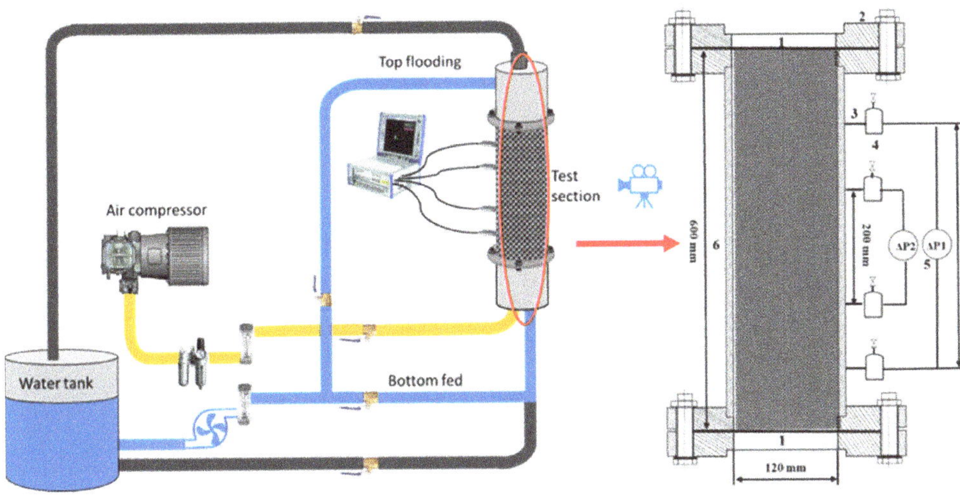

Figure 1. Diagram of DEBECO-LT and test section. 1—Stainless steel wire mesh, 2—Fittingflange, 3—Connecting pipe, 4—Gas–liquid separator, 5—Differential pressure transmitter, 6—Test section.

The experimental pressure is equal to the local atmospheric pressure, about 0.1 MPa. The room temperature is about 10 °C. Spherical glass particles with diameters of 1.5 mm, 3 mm, and 6 mm are packed into three different porous media beds. The superficial velocity of air is 0.005–0.44 m/s and that of water is 0.59–1.17 mm/s. More details of the experimental setup can be found in Table 1.

Table 1. Details of experiment setup.

Particle Sizes (mm)	Porosity	Superficial Velocity		Reynolds Number	
		Water (mm/s)	Gas (m/s)	Water	Gas
1.5	0.391	0.59~1.17	0.005–0.44	1.13–2.26	1.87–97.93
3	0.385	0.29~1.17	0.005–0.44	1.21–5.12	1.98–161.92
6	0.4	0.29~1.17	0.005–0.44	2.31–10.01	3.65–327.55

3. Methodology

3.1. Feature Extraction Methodology

Feature extraction is the key to establishing the relationship between flow patterns and signals, and it is also the basis of machine learning technology to distinguish and recognize flow patterns. Appropriate feature extraction is important to developing flow pattern identification technology. In this paper, statistical methods and empirical modal analysis (EMD) technology are used to extract the time domain and time-frequency domain characteristics of the signal.

3.1.1. Characteristics of Time Domain

In this paper, three common statistical parameters, which are mean, variance, and range of the signal, are selected to reflect the time domain characteristics of the signal. These features can be used to analyze the time domain features of signals and as the input of machine learning classifiers containing time domain features. The calculation formulas are as follows:

Mean of differential pressure signal:

$$m = \frac{1}{N}\sum_{i=1}^{N} s(i) \tag{1}$$

where $x(i)$ refers to each point of the signal, and N indicates the number of signal points.

The standard deviation of differential pressure signal:

$$S = \sqrt{\frac{1}{N}\sum_{i=1}^{N}(s(i) - m)^2} \tag{2}$$

where m refers to the mean of the differential pressure signal

Range of differential pressure signal:

$$R = s(i)_{\max} - s(i)_{\min} \tag{3}$$

3.1.2. Characteristics of Time-Frequency Domain

It is not comprehensive to extract the signal features only from the time domain. In order to extract more abundant features, this paper also introduces the empirical mode decomposition (EMD) method to extract the time-frequency domain features of the signal. EMD is a new adaptive signal time-frequency processing method proposed by N. E. Huang that is especially suitable for the analysis and processing of nonlinear and nonstationary signals. The essence of the empirical mode decomposition (EMD) method is to identify all vibration modes contained in the signal through the characteristic time scale. In this process, the characteristic time scale and the definition of IMF have certain experiences and approximations. Compared with other signal processing methods, the EMD method is intuitive, indirect, a posteriori, and adaptive. The characteristic time scale used in its decomposition is derived from the original signal.

The purpose of EMD decomposition is to obtain the eigenmode function. EMD decomposes the input signal into several eigenmode functions and a residual; the process is as follows:

(1) Find all extreme points of signal $I(n)$.
(2) Use a cubic spline curve to fit the envelope Emax(n) and Emin(n) of the upper and lower extreme points, and find the average value $m_1(n)$ of the upper and lower envelope and subtract it from $I(n)$:

$$h(n) = I(n) - m_1(n) \tag{4}$$

(3) Judge whether $h(n)$ is IMF according to preset criteria:

(1) In the whole time range of the function, the number of local extreme points and zero crossing points must be equal or be at most one difference;
(2) At any time point, the envelope of the local maximum (upper envelope) and the envelope of the local minimum (lower envelope) must be zero on average;

(4) If not, replace $I(n)$ with $h(n)$ and repeat the above steps until $h(n)$ meets the criteria; then $h(n)$ is the $IMF_m(n)$ to be extracted;
(5) Each time the $IMF_m(n)$ is obtained, it is deducted from the original signal and the above steps are repeated until the last remaining part $RES_M(n)$ of the signal is only a monotone sequence or a constant value sequence. In this way, the original signal $I(n)$ is decomposed into the linear superposition of a series of $IMF_m(n)$ and the remaining parts:

$$I(n) = \sum_{m=1}^{M} IMF_m(n) + RES_M(n) \tag{5}$$

where $I(n)$ indicates the input signal, $IMF_m(n)$ represents the eigenmode function of mth, and $RES_M(n)$ indicates the residual.

The EMD energy of mth is determined by the amplitude of $IMF_m(n)$, which is defined as follows:

$$E_m = \sum (IMF_m(n))^2 \tag{6}$$

The EMD energy spectrum is defined as follows:

$$E = \sum_{m=1}^{n} E_m \tag{7}$$

$$Level\ m = E_m / E \tag{8}$$

3.2. Machine Learn Methodologies

3.2.1. Support Vector Machine

Support vector machine (SVM) theory is proposed by Vapnik [28]. Generally speaking, the support vector machine is a classifier. For the two groups of marked vectors, an optimal segmentation hypersurface is given to divide the two groups of vectors into two sides so that the vector closest to the hyperplane in the two groups of vectors (the so-called support vector) is as far away from the hyperplane as possible. Figure 2 shows the schematic diagram of SVM.

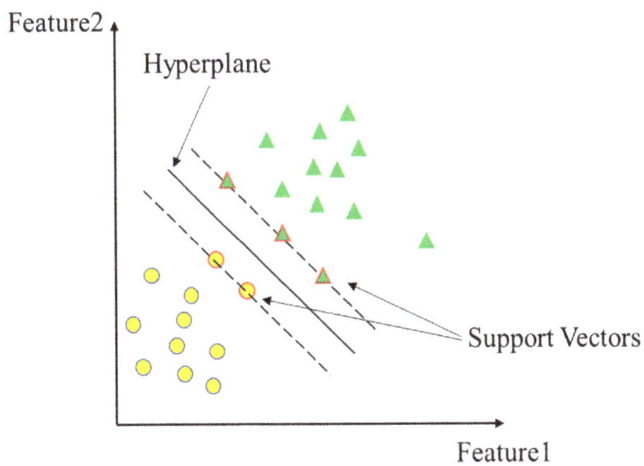

Figure 2. The classification principle of SVM.

For a data set, $data = \{(x_1, y_1), (x_2, y_2), \ldots, (x_n, y_n)\}$, where $x_i \in R^n, y_i \in \{+1, -1\}$. It can be specifically described as the following problem of finding the conditional maximum value:

$$\begin{cases} \min_a \left\{ \frac{1}{2} \sum_{i=1}^{m} \sum_{j=1}^{m} a_i a_j y_i y_j \phi(x_i)^T \phi(x_j) - \sum_{i=1}^{m} a_i \right\} \\ \sum_{j=1}^{m} a_i y_i = 0 \\ 0 \leq a_i \leq C \\ i = 1, 2, \ldots, m \end{cases} \tag{9}$$

The above problem involves the vector inner product mapped to a high-dimensional space, that is, the calculation of $\phi(x_i)^T \phi(x_j)$. Therefore, it is necessary to introduce kernel function $K(x,z)$. This paper employs the Gaussian kernel function:

$$K(x, z) = \exp\left(-\frac{\|x - z\|^2}{2\sigma^2} \right) \tag{10}$$

Besides, LIBSVM tools [29] provide support for the calculation.

3.2.2. BP Neural Network

BP (Back Propagation) neural network is a multilayer feedforward neural network trained according to the error backpropagation algorithm, which is one of the most widely used neural network models. The process of BP neural network is mainly divided into two stages. The first stage is the forward propagation of the signal, from the input layer to the hidden layer, and finally to the output layer. The second stage is the back-propagation of error, from the output layer to the hidden layer, and finally to the input layer, adjusting the weight and offset from the hidden layer to the output layer and from the input layer to the hidden layer. Figure 3 shows the schematic diagram of a BP neural network.

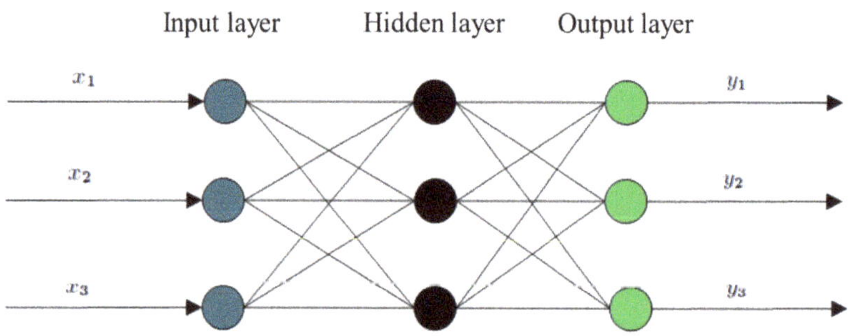

Figure 3. Schematic diagram of BP neural network.

4. Results and Discussion

4.1. Experiment Results

In this study, different flow patterns in a porous bed are recorded by changing the gas flow rate while the liquid flow rate is fixed. During this process, bubbly flow, slug flow, annular flow, and the transition flow patterns among them are observed. The photos and diagrams of typical flow patterns are shown in Figure 4. At first, the gas flow rate is lower. At this time, the gas phase is scattered between the pores of the porous media bed in the form of small bubbles. With the continuous increase in gas flow rate, the number and volume of bubbles increase, and the flow pattern begins to transition to slug flow. When the flow pattern is completely transformed into slug flow, it can be seen that the bubbles between adjacent pores are connected to form irregular gas slugs, as shown in Figure 4b. In the whole experimental section, such gas slugs can be seen everywhere. As the gas flow rate further increases, the gas slugs are connected to form larger gas slugs, and the flow pattern transitions to annular flow. For a typical annular flow, the liquid is attached to the surface of particles and pipe walls in the form of a liquid film, and the gas flows rapidly through the gap between these liquid films, as shown in Figure 4c. It is worth mentioning that in porous beds with particle diameters of 1.5 mm and 6 mm, three typical flow patterns cannot be completely observed within the current working conditions due to the limitation of pore size. Bubbly flow cannot be observed in a porous bed with a particle diameter of 1.5 mm. For a porous bed with a particle diameter of 6 mm, annular flow cannot be observed. Only in the porous bed with a particle diameter of 3 mm are bubbly flow, slug flow, and annular flow are observed at the same time.

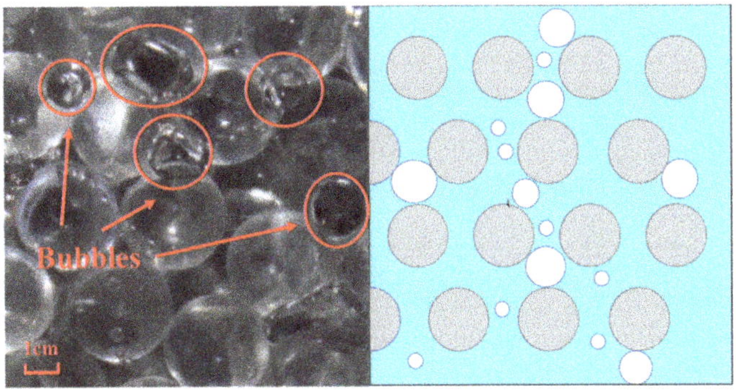

(**a**)

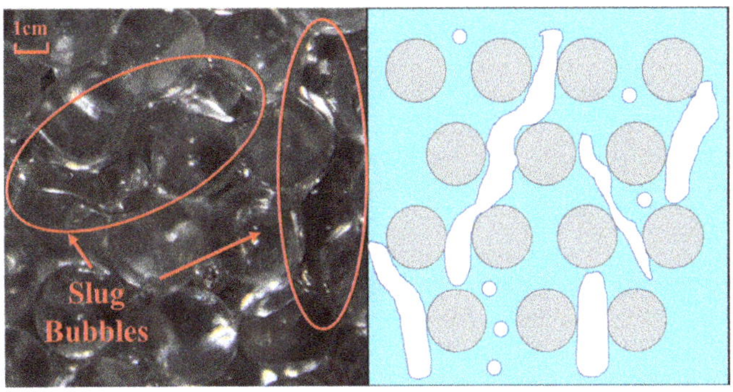

(**b**)

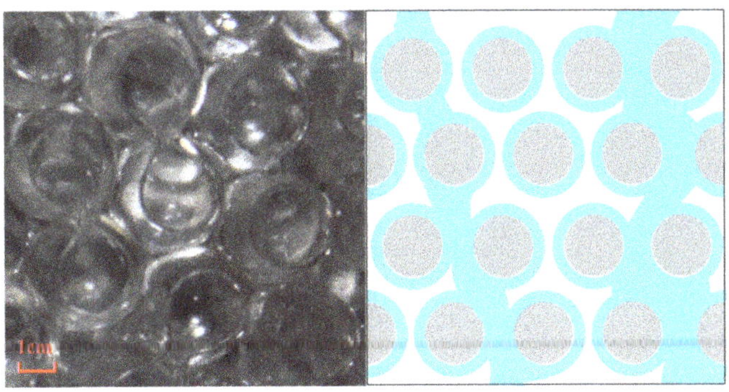

(**c**)

Figure 4. Photos and diagrams of typical flow patterns. (**a**). Photo and diagram of bubbly flow. (**b**). Photo and diagram of slug flow. (**c**). Photo and diagram of annular flow.

Figure 5 shows the diagram of differential pressure signals of typical flow patterns in porous beds with different particle diameters. As shown in Figure 5a,b, the differential pressure signals of bubbly flow in two porous beds fluctuate around 4.5 kPa, and the fluctuation ranges are about 1 kPa. The difference in particle diameter does not lead to an obvious difference between differential pressure signals. For slug flow, it can be seen from Figure 5c–e that there is a great difference between the differential pressure signals of bubbly flow and that of slug flow. The differential pressure signals of slug flow in porous beds with particle diameters of 3 mm and 1.5 mm fluctuate up and down around 5 kPa, and the fluctuation ranges can reach about 2 kPa, which is larger than that of bubbly flow. As for the differential pressure signal of slug flow in a porous bed with a particle diameter of 6 mm, although its fluctuation range is small, which is about 1.5 kPa. The average of the signal is around 3.5 kpa at this time, which is significantly smaller than 4.5 kPa of bubbly flow and 5 kPa of slug flow in the other two porous beds. By observing differential pressure signals of annular flow in Figure 5f,g, it can be found that the average value of the differential pressure signals is increased, both greater than 5 kPa. For the porous bed with a particle diameter of 3 mm, the fluctuation range of the differential pressure signal of annular flow is smaller than that of the slug flow.

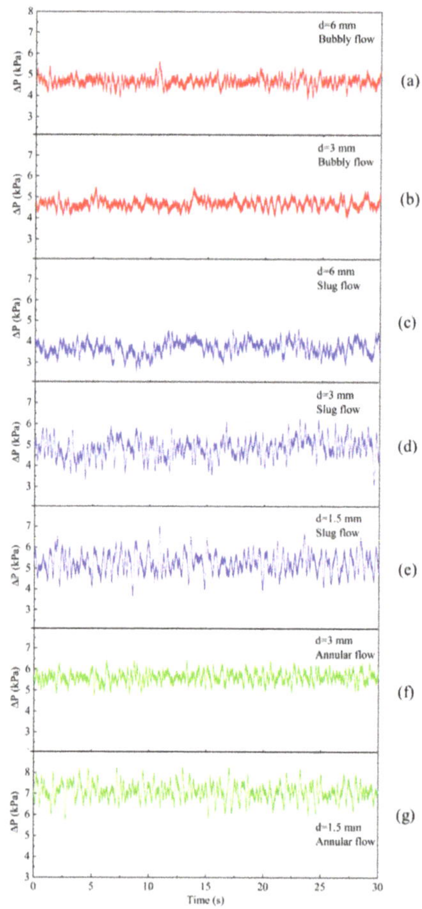

Figure 5. Diagram of differential pressure signals of typical flow patterns in different porous beds. (**a**). d = 6 mm, bubbly flow. (**b**). d = 3 mm, bubbly flow. (**c**). d = 6 mm, slug flow. (**d**). d = 3 mm, slug flow. (**e**). d = 1.5 mm, slug flow. (**f**). d = 3 mm, annular flow. (**g**). d = 1.5 mm, annular flow.

In general, it can be found that there are certain differences in the features of differential pressure signals of different flow patterns, such as the mean value, dispersion degree, and so on. The features of these typical flow patterns are extracted, quantified, and classified, and then the unknown flow patterns can be classified and recognized through these features. The next section will introduce how to use time domain analysis and EMD methods to extract and quantify the characteristic parameters of differential pressure signals.

4.2. Feature Extraction

4.2.1. Time Domain Feature Extraction

According to Equations (1)–(3), three time domain parameters (average value, standard deviation, and range) corresponding to each group of data are obtained. The average reflects the concentration trend of the differential pressure signal, the standard deviation reflects the dispersion degree of the differential pressure signal, and the range quantifies the fluctuation range of the differential pressure signal. After sorting out the calculation results, the time domain characteristic distributions of different flow patterns in porous beds with different particle diameters are obtained, as shown in Figure 6.

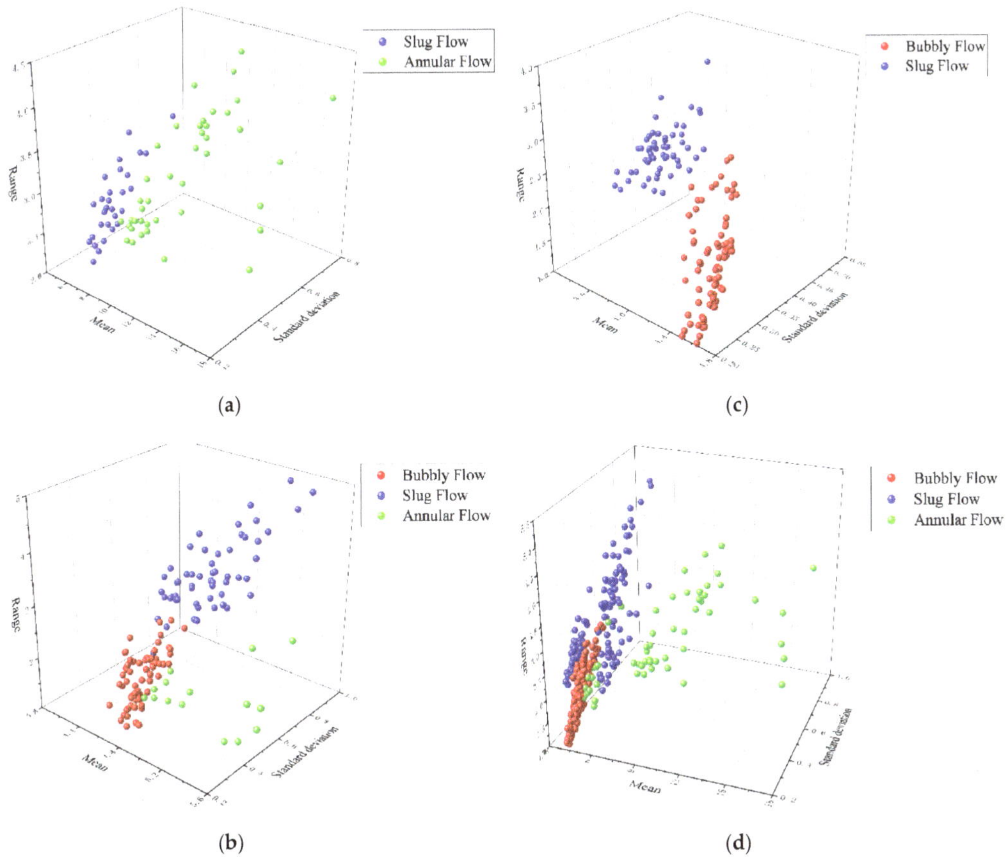

Figure 6. Diagram of time domain characteristic parameter distribution. (**a**). Time domain characteristic parameter distribution in porous bed with a particle diameter of 1.5 mm. (**b**). Time domain characteristic parameter distribution in porous bed with a particle diameter of 3 mm. (**c**). Time domain characteristic parameter distribution in porous bed with a particle diameter of 6 mm. (**d**). Time domain characteristic parameter distribution in all porous beds.

It can be seen from Figure 6a that there is an obvious boundary between the slug flow and annular flow in a porous bed with a particle diameter of 1.5 mm. Figure 6b shows that in a porous bed with a particle diameter of 3 mm, a few data points of slug flow and bubble flow overlap in the time domain distribution. In Figure 6c, it can be seen that in the porous bed with a particle diameter of 6 mm, there is also an obvious boundary between the distribution of bubble flow and slug flow data in the time domain. Nevertheless, it can be found from Figure 6d that when the influence of particle diameter of porous media bed is not considered, there will be a lot of overlap in the distribution of time domain characteristics of different flow patterns, which indicates that particle diameter is also one of the important parameters affecting classification, and the influence of particle diameter cannot be ignored.

In order to better analyze the distribution of time domain characteristics, the distribution range and median of time domain characteristic parameters of different flow patterns are counted, and the calculation results are shown in Table 2.

Table 2. Distribution of time domain parameters of flow patterns.

Particle Diameter	Quartile	Bubbly Flow			Slug Flow			Annular Flow		
		m	*S*	*R*	*m*	*S*	*R*	*m*	*S*	*R*
1.5 mm	min				5.74	0.28	2.08	8.02	0.33	2.35
	median				6.13	0.38	2.65	10.27	0.46	3.11
	max				7.05	0.60	3.50	17.48	0.74	4.40
3 mm	min	4.65	0.21	1.47	4.73	0.38	2.43	4.93	0.27	1.79
	median	4.85	0.30	2.18	5.05	0.54	3.64	5.27	0.29	2.26
	max	5.01	0.44	3.01	5.49	0.86	5.08	5.88	0.51	3.25
6 mm	min	4.11	0.17	1.26	3.38	0.31	1.99			
	median	4.60	0.26	1.91	3.61	0.37	2.52			
	max	4.76	0.40	2.71	3.97	0.50	3.51			
all	min	4.11	0.17	1.26	3.38	0.28	1.99	4.93	0.27	1.79
	median	4.70	0.28	1.98	4.91	0.41	2.79	9.00	0.40	2.72
	max	5.01	0.44	3.01	7.05	0.86	5.08	17.48	0.74	4.40

It can be seen from the Table 2 that for the porous bed with a particle diameter of 1.5 mm, the average value of the differential pressure signal of the slug flow is 5.74–7.05 kPa, and that of the annular flow is greater than 8.02 kPa. Therefore, the data with the average value of the differential pressure signal falling within these ranges can be directly distinguished between the slug flow and the annular flow in the porous bed with a particle diameter of 1.5 mm. Similarly, in the porous media bed with a particle diameter of 6 mm, the average value of the differential pressure signal can also be directly used to distinguish bubbly flow and slug flow (bubbly flow is 4.11–4.76 kPa, slug flow is 3.38–3.97 kPa). However, it should be considered that the average value of differential pressure signal alone cannot distinguish all the flow patterns. Firstly, in a porous media bed with a particle diameter of 3 mm, there is a range of overlap between the characteristic parameters of different flow patterns. Secondly, for porous media beds with particle diameters of 1.5 and 6 mm, it is also difficult to distinguish data that are between two ranges (for example, the average value of differential pressure signal is within 7.05–8.02 kPa) through the judgment of researchers. This is why multiple parameters rather than a single parameter are selected to judge the flow pattern. To solve the above problems, new features (EMD energy spectrum) are introduced to provide a more judgment basis. At the same time, machine learning technology is employed to identify and classify different flow patterns from a mathematical point of view.

4.2.2. Time-Frequency Feature Extraction

EMD has been briefly introduced in Section 3.1.2. In this section, the differential pressure signals of different flow patterns are decomposed to obtain the IMF components of the signal; the decomposition results are shown in Figure 7.

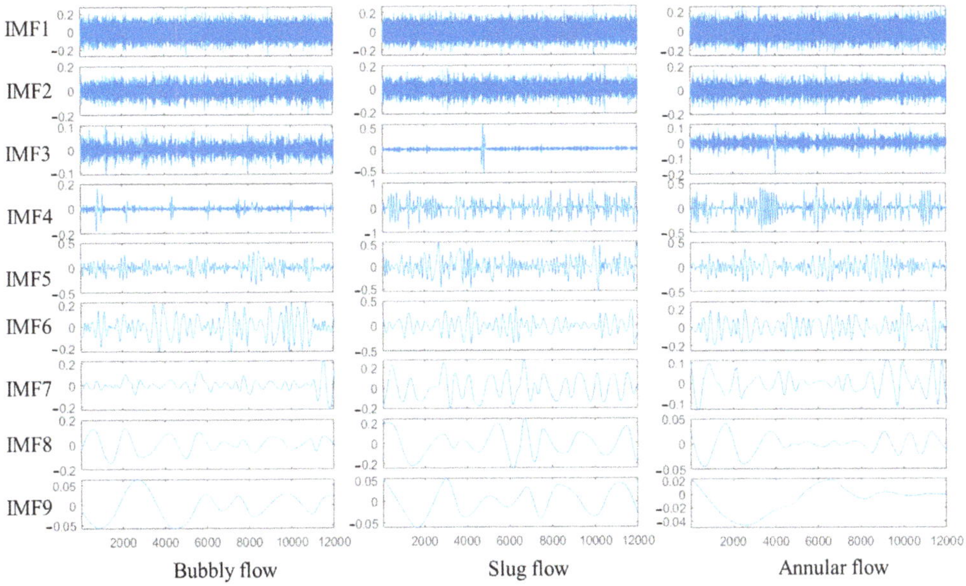

Figure 7. IMF components of different flow patterns.

As can be seen in Figure 7, most of the IMF3 of slug flow is close to zero, which is obviously different from that of bubbly flow and annular flow. Similarly, the IMF4 of bubbly flow is also very different from that of slug flow and annular flow, not only in forms but also in amplitude. Moreover, for IMF6, IMF7, there are also differences in the amplitudes of modal functions of different flow patterns. Therefore, using Equations (6) and (7), the EMD energy spectra of different signals are defined and calculated, and the corresponding distribution ranges are obtained. Due to the adaptability of the EMD method, the decomposition levels of different signals are different. Considering that the amplitudes of IMF8 and IMF9 of most signals are very low (energy is very low), in order to unify the results, only the energy of level one–level seven is retained. The calculation results are shown in Tables 3–5. It can be seen from Table 3 that the energy proportion of EMD at levels 1–3 is not high, especially from the median point of view; only the bubbly flow in a porous bed with a particle diameter of 6 mm accounts for more than 10% of the energy at level one. It can be seen from Table 4 that the EMD energy of the signal is mainly distributed at levels four, five, and six, especially at level four. As can be seen in Table 5, the difference in EMD energy at level seven between different conditions is large. For example, for the slug flow in the porous bed with particle diameters of 3 mm and 6 mm, the energy proportion may reach more than 40%, while for other conditions, the energy proportion of this level is much lower.

In general, by processing the signal with the EMD method and calculating the corresponding EMD energy spectrum, new parameters that are different from time domain characteristic parameters can be obtained. These new parameters will show different characteristics with different flow patterns, which can be used for the identification of flow patterns. They also provide richer feature vectors as support for the next machine learning classification.

Table 3. Distributions of EMD energy spectrum on levels 1–3.

Particle Diameter	Quartile	Bubbly Flow			Slug Flow			Annular Flow		
		Level 1	Level 2	Level 3	Level 1	Level 2	Level 3	Level 1	Level 2	Level 3
1.5 mm	min				2.38%	1.15%	0.35%	1.46%	0.74%	0.29%
	median				5.50%	2.55%	0.86%	3.83%	1.81%	3.23%
	max				9.06%	4.37%	8.97%	7.61%	3.58%	30.81%
3 mm	min	3.05%	1.53%	0.48%	1.05%	0.54%	0.21%	3.07%	1.47%	0.40%
	median	7.55%	3.53%	0.97%	2.48%	1.23%	1.94%	8.84%	4.15%	1.53%
	max	16.77%	7.71%	7.02%	5.01%	2.31%	12.42%	12.24%	5.88%	5.96%
6 mm	min	4.41%	2.34%	0.78%	2.81%	1.51%	0.56%			
	median	10.36%	5.25%	1.46%	5.24%	2.86%	0.90%			
	max	26.74%	12.76%	3.55%	7.50%	4.04%	2.29%			

Table 4. Distributions of EMD energy spectrum on levels 4–6.

Particle Diameter	Quartile	Bubbly Flow			Slug Flow			Annular Flow		
		Level 4	Level 5	Level 6	Level 4	Level 5	Level 6	Level 4	Level 5	Level 6
1.5 mm	min				12.89%	5.62%	4.17%	14.67%	8.20%	3.18%
	median				41.20%	15.76%	8.70%	49.53%	19.06%	6.34%
	max				54.80%	28.95%	15.18%	63.68%	32.43%	15.24%
3 mm	min	4.20%	7.78%	6.28%	18.91%	8.80%	3.56%	17.22%	9.26%	2.95%
	median	27.93%	23.54%	15.71%	38.60%	16.42%	9.37%	47.36%	25.62%	5.35%
	max	41.96%	36.65%	30.57%	54.29%	31.79%	23.24%	53.11%	36.10%	9.27%
6 mm	min	15.70%	11.03%	6.56%	8.97%	10.90%	5.52%			
	median	32.78%	22.00%	13.26%	22.36%	20.97%	15.36%			
	max	46.04%	37.71%	27.43%	32.97%	33.70%	31.20%			

Table 5. Distributions of EMD energy spectrum on level 7.

Particle Diameter	Quartile	Bubbly Flow	Slug Flow	Annular Flow
		Level 7	Level 7	Level 7
1.5 mm	min		2.80%	0.52%
	median		7.88%	3.38%
	max		15.33%	13.96%
3 mm	min	3.71%	1.40%	1.19%
	median	10.94%	12.61%	2.13%
	max	24.44%	46.50%	9.21%
6 mm	min	1.65%	3.51%	
	median	5.02%	14.75%	
	max	12.52%	43.14%	

4.3. Machine Learning Identification

In Section 4.2, the time domain characteristics and EMD energy spectrum of differential pressure signals are calculated. After analyzing these characteristics, it can be found that it is difficult to classify and identify all working conditions with a single parameter, and there is often overlap between parameters. The complex characteristics make it difficult for researchers to accurately identify flow patterns through subjective judgment. The development of machine learning technology, especially support vector machine (SVM) technology and neural network technology, provides new ideas and tools for solving these problems. Through SVM technology and neural network technology, the recognition model can be trained by using multi-dimensional feature vectors based on certain mathematical rules. Compared with manual recognition, these mathematical models can quickly and accurately judge the flow pattern according to the input feature vector, which greatly reduces the impact of subjectivity on the results and improves the recognition efficiency. Therefore, this section uses SVM technology and neural network technology to train multiple SVM/neural network identification models with different features and compares the identification ability of these models to evaluate the performance of the two technologies in porous media flow pattern identification.

Samples should be prepared before training. A total of 341 sets of typical flow pattern data are obtained in this experiment, including bubbly flow, slug flow, and annular flow. One part of the data is used as the training set to train the model, and the other part of the data is used as the test set to test the identification ability of the model. The details are shown in Table 6.

Table 6. The number of samples of different flow patterns.

	Flow Pattern Data Sets		
	Train	**Test**	**Total**
Bubbly	77	63	140
Slug	80	65	145
Annular	31	25	56
total	188	153	341

The vector including time domain features and EMD energy spectrum is constructed. Considering the important influence of particle diameter on the characteristics, the particle diameter is also included in the vector. The dimension n of the vector should not be too large, which is limited by the number of training sets. In this paper, $n < 8$ ($2^n < 188$). Therefore, the final vector is composed of three time domain characteristic parameters, three levels of EMD energy ratio, and one particle diameter. From large to small, two kinds of vectors are constructed, which will lead to two different SVM models. One vector includes four, five, and six levels and the other one includes one, two, and seven levels. Examples of vectors are shown in Table 7.

Table 7. Examples of vectors with different features.

Vector Type	Parameters							Label	Flow Pattern
	m	S	R	Level 4	Level 5	Level 6	d_p		
Vector-1	4.86	0.22	1.64	12.41%	16.73%	30.57%	3	1	Bubbly
	3.96	0.39	2.69	22.55%	24.26%	15.26%	6	2	Slug
	8.74	0.34	2.71	55.92%	12.33%	9.04%	1.5	3	Annular
	m	S	R	Level 1	Level 2	Level 7	d_p		
Vector-2	4.86	0.22	1.64	16.24%	7.14%	12.95%	3	1	Bubbly
	3.96	0.39	2.69	3.60%	1.95%	13.84%	6	2	Slug
	9.08	0.34	2.72	6.23%	2.85%	2.90%	1.5	3	Annular

The SVM models are trained with the prepared vectors. In this paper, the k-CV method [30] combined with the grid search method is used to obtain two key parameters (the penalty factor C and the kernel function parameter g), which are important to ensure the accuracy of SVM models. Table 8 shows the identification results of the two SVM models.

In general, the identification ability of the SVM-1 model is better than that of the SVM-2 model. The identification ability of slug flow and annular flow of SVM-1 model is better than that of the SVM-2 model. However, its bubbly flow identification ability is worse than that of the SVM-2 model, but this gap is not very obvious.

In addition to the SVM model, the neural network flow pattern identification model is also trained based on BP neural network technology. In this paper, BP neural network includes one input layer (7 neurons, corresponding to seven features of a vector), one intermediate layer (12 neurons), and one output layer (3 neurons, corresponding to labels of three flow patterns). The same training set and test set as SVM models are selected, which can help evaluate the identification ability of the two methods. The training results are shown in Table 9.

Table 8. The identification results of SVM models.

SVM Model	Flow Pattern	Correct Identification	Total Number	Accuracy
SVM-1	Bubbly	60	63	95.24%
	Slug	62	65	95.38%
	Annular	23	25	92.00%
	Overall	145	153	94.77%
SVM-2	Bubbly	61	63	96.83%
	Slug	60	65	92.31%
	Annular	22	25	88.00%
	Overall	143	153	93.46%

Table 9. The identification results of BP-network models.

BP-Network Model	Flow Pattern	Correct Identification	Total Number	Accuracy
BP-1	Bubbly	61	63	96.83%
	Slug	63	65	96.92%
	Annular	23	25	92.00%
	Overall	147	153	96.08%
BP-2	Bubbly	58	63	92.06%
	Slug	61	65	93.85%
	Annular	21	25	84.00%
	Overall	140	153	91.50%

As shown in Table 9, the identification ability of different BP network models varies greatly. The BP-1 model has the best identification ability, which is not only better than the BP-2 model but also better than SVM-1 and SVM-2 models. The identification ability of the BP-2 model is the worst of the four models. Moreover, the models containing one, two, and seven levels of EMD energy perform worse than the models containing four, five, and six levels of EMD energy.

Next, consider further improving the identification ability of flow patterns. The idea of bagging technology in integrated learning technology is introduced. That is, train multiple models at the same time, let these models vote on the results, and determine the final results according to the voting results. In this study, four models are trained. If all the models participate in the voting, there may be a tie vote. Due to the poor performance of BP-2, its results may mislead the correct results. Therefore, SVM-1, SVM-2, and BP-1 were selected to form an integrated model. The identification process is shown in Figure 8, moreover, considering the special situation (bubble flow: annular flow: slug flow, 1:1:1). Since the BP-1 model performed best among the three models, it is considered that the final result of BP-1 model will prevail when the conflict occurs. Table 10 shows the identification results of an integrated model. It can be seen that the integrated model further improves identification ability based on the original models. The test data of bubbly flow and annular flow have been correctly identified, and only two groups of slug flow data have not been correctly identified.

Table 10. Identification results of integrated model.

	Flow Pattern	Correct Identification	Total Number	Accuracy
Integrated model	Bubbly	63	63	100.00%
	Slug	63	65	96.92%
	Annular	25	25	100.00%
	Overall	151	153	98.69%

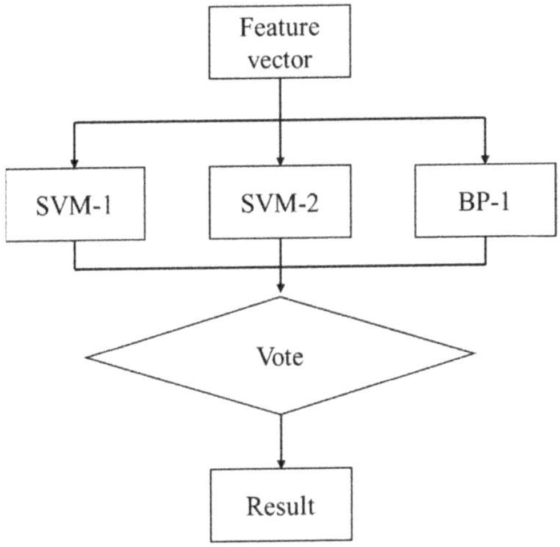

Figure 8. Diagram of integrated model.

5. Conclusions

In this paper, a method based on feature extraction and machine learning is proposed to identify the two-phase flow patterns in porous media, which provides a new idea for the development of flow pattern identification methods in porous media in the fields of the chemical industry, agriculture, petroleum, and nuclear engineering. The differential pressure signals of two phase flow in the porous media packed with particles are collected through visual experiments. The time domain characteristic parameters and EMD energy spectrum are extracted. After that, a variety of flow pattern identification models based on machine learning technology are trained. An integrated flow pattern identification model with high accuracy is finally obtained based on integrated identification technology. The main conclusions are as follows:

(1) There are differences between the average values of differential pressure signals of different flow patterns. For a porous bed with a particle diameter of 1.5 mm, the boundary between slug flow and annular flow is 8 kPa. For a porous bed with a particle diameter of 3 mm, the distribution range of average differential pressure of different flow patterns overlaps. For a porous bed with a particle diameter of 6 mm, the boundary between bubbly flow and slug flow is 4 kPa. For other parameters, the overlapping area between different flow patterns is larger, and it is difficult to distinguish the flow patterns only by manual identification. It is necessary to introduce machine learning technology.

(2) The BP-1 model based on BP neural network technology has the best identification ability among single models, with an accuracy of 96.08%. However, another BP-2 model based on different levels of EMD energy has the worst identification ability. Its accuracy is only 91.5%. The identification ability of the two models SVM-1 and SVM-2 trained by SVM technology is close, since the accuracies of them are 94.77%, and 93.4%, respectively. In this study, the two neural network models have the highest and lowest recognition accuracy. Although the SVM model is lower than the optimal neural network model in recognition accuracy, the recognition ability of the two models is closer. SVM technology is more stable than BP neural network technology.

(3) By integrating several high-quality models, the integrated model can further improve the ability of flow pattern identification on the basis of the original models. The identification accuracy increased from 94.77% to 98.04%. This behavior will increase the total calculation time because it takes time to train each model. The total time is approximately equal to the sum of the time needed to train the three models,

respectively. Users can consider comprehensively according to the requirements for accuracy and timeliness. Moreover, poor quality models will reduce the identification ability of integrated models.

Author Contributions: Conceptualization, L.L.; Data curation, X.L. and J.Z.; Formal analysis, W.W., H.Z. and J.Z.; Investigation, X.L. and L.L.; Methodology, X.L., L.L. and J.Z.; Software, W.W. and H.Z.; Visualization, W.W. and H.Z.; Writing—original draft, X.L.; Writing—review & editing, L.L. All authors have read and agreed to the published version of the manuscript.

Funding: This research received no external funding.

Conflicts of Interest: The authors declare no conflict of interest.

Nomenclature

a	approximation coefficient, 1
b	Constant term of hyperplane equation,1
C	Crest factor, 1
E	local energy density, 1
J	flow rate, mm/s
K	kernel function
P	power spectral density, kPa^2/Hz
R	Range value
s	signal
S	Standard deviation
t	time
x	Coordinates
y	Coordinates
z	Coordinates
ϕ	scaling function
Subscripts	
g	gas
i	unit number
j	unit number
k	unit number
m	unit number
max	maximum
min	minimum
M	unit number
R	Range value
S	Standard deviation
Acronyms	
BP	Back propagation
EMD	Empirical mode decomposition
IMF	Intrinsic mode function
SVM	support vector machine

References

1. Ma, W.; Yuan, Y.; Sehgal, B.R. In-Vessel Melt Retention of Pressurized Water Reactors: Historical Review and Future Research Needs. *Engineering* **2016**, *2*, 103–111. [CrossRef]
2. Hu, B.; Gu, Z.; Zhou, C.; Wang, L.; Huang, C.; Su, J. Investigation of the Effect of Capillary Barrier on Water–Oil Movement in Water Flooding. *Appl. Sci.* **2022**, *12*, 6285. [CrossRef]
3. Kulli, B. Visualizing soil compaction based on flow pattern analysis. *Soil Tillage Res.* **2003**, *70*, 29–40. [CrossRef]
4. Zhang, C.; Jiao, W.; Liu, Y.; Qi, G.; Yuan, Z.; Zhang, Q. CFD Simulation of Dry Pressure Drop in a Cross-Flow Rotating Packed Bed. *Appl. Sci.* **2021**, *11*, 10099. [CrossRef]
5. Yao, T.; Wu, Q.; Liu, Z.; Zou, S.; Xu, Q.; Guo, L. Experimental investigation on mitigation of severe slugging in pipeline-riser system by quasi-plane helical pipe device. *Exp. Therm. Fluid Sci.* **2019**, *102*, 189–204. [CrossRef]
6. Tosun, G. A study of cocurrent downflow of nonfoaming gas-liquid systems in a packed bed. 1. Flow regimes: Search for a generalized flow map. *Ind. Eng. Chem. Process Des. Dev.* **1984**, *23*, 29–35. [CrossRef]

7. Tung, V.; Dhir, V. A hydrodynamic model for two-phase flow through porous media. *Int. J. Multiph. Flow* **1988**, *14*, 47–65. [CrossRef]
8. Xu, G.; Zhang, X.; Sun, Z.; Ruan, J.; He, B. Flow patterns and transition criteria in boiling water-cooled packed bed reactors. *Prog. Nucl. Energy* **2018**, *108*, 214–221. [CrossRef]
9. Shaban, H.; Tavoularis, S. Identification of flow regime in vertical upward air–water pipe flow using differential pressure signals and elastic maps. *Int. J. Multiph. Flow* **2014**, *61*, 62–72. [CrossRef]
10. Wu, B.; Ribeiro, A.S.; Firouzi, M.; Rufford, T.E.; Towler, B. Use of pressure signal analysis to characterise counter-current two-phase flow regimes in annuli. *Chem. Eng. Res. Des.* **2020**, *153*, 547–561. [CrossRef]
11. Vieira, S.C.; van der Geest, C.; Fabro, A.; de Castro, M.S.; Bannwart, A.C. Intermittent slug flow identification and characterization from pressure signature. *Mech. Syst. Signal Process.* **2021**, *148*, 107148. [CrossRef]
12. Guo, W.; Liu, C.; Wang, L. Temperature fluctuation on pipe wall induced by gas–liquid flow and its application in flow pattern identification. *Chem. Eng. Sci.* **2021**, *237*, 116568. [CrossRef]
13. Huang, N.E.; Shen, Z.; Long, S.R.; Wu, M.C.; Shih, H.H.; Zheng, Q.; Yen, N.-C.; Tung, C.C.; Liu, H.H. The empirical mode decomposition and the Hilbert spectrum for nonlinear and non-stationary time series analysis. *Proc. R. Soc. Lond. Ser. A Math. Phys. Eng. Sci.* **1998**, *454*, 903–995. [CrossRef]
14. Matsui, G. Identification of flow regimes in vertical gas-liquid two-phase flow using differential pressure fluctuations. *Int. J. Multiph. Flow* **1984**, *10*, 711–719. [CrossRef]
15. Elperin, T.; Klochko, M. Flow regime identification in a two-phase flow using wavelet transform. *Exp. Fluids* **2002**, *32*, 674–682. [CrossRef]
16. dos Reis, E.; Goldstein, L., Jr. Characterization of slug flows in horizontal piping by signal analysis from a capacitive probe. *Flow Meas. Instrum.* **2010**, *21*, 347–355. [CrossRef]
17. Li, X.; Wei, T.; Wang, D.; Hu, H.; Kong, L.; Xiang, W. Study of gas–liquid two-phase flow patterns of self-excited dust scrubbers. *Chem. Eng. Sci.* **2016**, *151*, 79–92. [CrossRef]
18. Khorasani, M.; Ghasemi, A.; Leary, M.; Sharabian, E.; Cordova, L.; Gibson, I.; Downing, D.; Bateman, S.; Brandt, M.; Rolfe, B. The effect of absorption ratio on meltpool features in laser-based powder bed fusion of IN718. *Opt. Laser Technol.* **2022**, *153*, 108263. [CrossRef]
19. Sezer, H.; Tang, J.; Ahsan, A.N.; Kaul, S. Modeling residual thermal stresses in layer-by-layer formation of direct metal laser sintering process for different scanning patterns for 316L stainless steel. *Rapid Prototyp. J.* **2022**. *ahead of print*. [CrossRef]
20. Rashed, K.; Kafi, A.; Simons, R.; Bateman, S. Fused Filament Fabrication of Nylon 6/66 Copolymer: Parametric Study Com-paring Full Factorial and Taguchi Design of Experiments. *Rapid Prototyp. J.* **2022**, *28*, 1111–1128. [CrossRef]
21. Agrawal, R. Sustainable design guidelines for additive manufacturing applications. *Rapid Prototyp. J.* **2022**, *28*, 1221–1240. [CrossRef]
22. Liang, F.; Hang, Y.; Yu, H.; Gao, J. Identification of gas-liquid two-phase flow patterns in a horizontal pipe based on ultrasonic echoes and RBF neural network. *Flow Meas. Instrum.* **2021**, *79*, 101960. [CrossRef]
23. Pei, S.; Liu, H.; Zhu, Y.; Zhang, C.; Zhao, M.; Fu, G.; Yang, K.; Yuan, Y.; Zhang, C. Identifying Flow Patterns in Water Pipelines Using Complex Network Theory. *J. Hydraul. Eng.* **2021**, *147*, 04021019. [CrossRef]
24. Zhang, L.; Wang, H. Identification of oil–gas two-phase flow pattern based on SVM and electrical capacitance tomography technique. *Flow Meas. Instrum.* **2010**, *21*, 20–24. [CrossRef]
25. Liu, W.; Tan, C.; Dong, F. Doppler spectrum analysis and flow pattern identification of oil-water two-phase flow using dual-modality sensor. *Flow Meas. Instrum.* **2021**, *77*, 101861. [CrossRef]
26. Ambrosio, J.D.S.; Lazzaretti, A.E.; Pipa, D.R.; da Silva, M.J. Two-phase flow pattern classification based on void fraction time series and machine learning. *Flow Meas. Instrum.* **2021**, *83*, 102084. [CrossRef]
27. Li, L.; Zou, X.; Wang, H.; Zhang, S.; Wang, K. Investigations on two-phase flow resistances and its model modifications in a packed bed. *Int. J. Multiph. Flow* **2018**, *101*, 24–34. [CrossRef]
28. Vapnik, V.N. An overview of statistical learning theory. *IEEE Trans. Neural Netw.* **1999**, *10*, 988–999. [CrossRef]
29. Chang, C.; Lin, C. LIBSVM: A Library for Support Vector Machines. *ACM Trans. Intell. Syst. Technol.* **2011**, *2*, 1–27. [CrossRef]
30. Rodriguez, J.D.; Perez, A.; Lozano, J.A. Sensitivity Analysis of k-Fold Cross Validation in Prediction Error Estimation. *IEEE Trans. Pattern Anal. Mach. Intell.* **2010**, 32, 569–575. [CrossRef]

applied sciences

MDPI

Article

Enhanced Heat Transfer Study of Spherical Heat Storage Based on Response Surface Methodology

Liwei Lu [1], Rui Tian [1,2,*], Xuan Gong [3] and Yuanxing Zhao [4]

1 College of Energy and Power Engineering, Inner Mongolia University of Technology, Hohhot 010051, China; 20171000025@imut.edu.cn
2 Inner Mongolia Key Laboratory of Renewable Energy, Hohhot 010051, China
3 College of Electrical Engineering, Inner Mongolia University of Technology, Hohhot 010051, China; gongxn2022@imut.edu.cn
4 College of Energy and Transportation Engineering, Inner Mongolia Agricultural University, Hohhot 010018, China; zhaoyx@imau.edu.cn
* Correspondence: tianr@imut.edu.cn

Abstract: In this paper, the effect of melting characteristics of CuO/paraffin wax composite phase change material in a spherical heat storage unit in a constant temperature water bath is investigated. Experiments were conducted in three different water bath temperatures (65 °C, 70 °C, and 75 °C). The inner surface of the sphere was fixed with two, four, and six pin-shaped fins 3 mm in diameter. The spheres were filled with different mass fractions of CuO nanoparticles/paraffin phase change materials. Experimental CCD was used to model and optimize the spherical thermal storage unit. Regression models were developed to predict the effects of various operational factors on the melting time of the composite PCM. The factors in the model included the number of pin fins in the spherical heat storage unit, the water bath temperature, and the content of added CuO nanoparticles in the PCM, and ANOVA was used to statistically validate the regression model. The results showed that the interaction between the water bath temperature and the number of pin fins had the most significant effect on the melting time. With the melting time of the phase change material as the optimized objective function, the optimized optimal working condition was six pin fins, a water bath temperature of 75 °C, and the addition of 5 wt% CuO nanoparticles/paraffin phase change material, and the actual melting time under this condition was 78.9 min, which was lower than the predicted value of 79.4 min, with an error of 0.63% between them.

Keywords: phase change material; spherical heat storage unit; fin; response surface methodology

Citation: Lu, L.; Tian, R.; Gong, X.; Zhao, Y. Enhanced Heat Transfer Study of Spherical Heat Storage Based on Response Surface Methodology. *Appl. Sci.* **2023**, *13*, 8595. https://doi.org/10.3390/app13158595

Academic Editor: Liangxing Li

Received: 25 June 2023
Revised: 20 July 2023
Accepted: 20 July 2023
Published: 26 July 2023

1. Introduction

In recent years, phase change thermal storage technology has been used in areas such as waste heat recovery and solar energy storage to cover the differences in quantity, form, and space between energy supply and demand. Phase change thermal storage materials are widely used in latent heat storage because of their high heat storage density, almost constant temperature during the heat storage/exothermic process, wide melting point distribution, chemical stability, and corrosion resistance [1]. Paraffin waxes are stable as phase change materials, free from subcooling and phase separation, cheap, and non-toxic, but they also have drawbacks, such as low thermal conductivity. With the rapid development of nanotechnology, the addition of nanoparticles (e.g., CuO [2,3], Al_2O_3 [4,5], Fe_3O_4 [6,7], SiO_2 [8,9], etc.) with high thermal conductivity to paraffin wax has become an emerging means [10]. Chen [11] prepared CuO/paraffin nanopowder composite phase change materials with mass fractions of 0.01–0.1%. The results showed that the addition of CuO nanopowder to paraffin wax could greatly improve the solar thermal conversion capacity by enhancing the light absorption capacity of the PCM. At low mass concentrations (0–0.1%), the steady-state temperature gradually increased with increasing mass fraction of

CuO nanoparticles, with a maximum increase of about 2.3 times. A nanocomposite for energy storage was prepared by Lin [12] by adding 20 nm nanoparticles of Cu to paraffin wax. The addition of 2.0% Cu nanoparticles to paraffin wax was found to increase the thermal conductivity of the Cu/paraffin phase change material by 46.3%. Kumar [13] dispersed ZnO nanoparticles in paraffin and analyzed their thermal properties. The experimental results showed that ZnO nanoparticles distributed inside the paraffin wax without affecting its chemical structure greatly improved its thermal stability and increased its thermal conductivity to 41.67% at 2.0 wt%. Pasupathi [14] experimentally investigated the effect of hybrid nanoparticles containing SiO_2 and CeO_2 nanoparticles on the thermophysical properties of paraffin-based phase change materials (PCMs). The experimental results showed that the mixed nanoparticles spread uniformly in the paraffin matrix without affecting the chemical arrangement of paraffin molecules. When hybrid nanoparticles were dispersed in paraffin, the relative thermal stability and relative thermal conductivity of paraffin were increased synergically, reaching 115.49% and 165.56%, respectively. In addition, the hybrid nanoparticles appropriately changed the melting point and crystallization point of the paraffin wax and reduced the degree of supercooling of the paraffin wax; the maximum degree of supercooling was reduced by 35.81%. Li [15] conducted an experimental study of the melting and solidification processes and the enhancement of thermal conductivity of PCM inside a sphere. The enhancement of the thermal conductivity of PCM by aluminum powder was investigated via homogeneous diffusion and sedimentation methods. The results showed that the addition of aluminum powder accelerated the melting and solidification processes of PCM within the sphere. However, the settling of the aluminum powder during the melting process was more conducive to accelerated heat transfer throughout the sphere than the uniform diffusion of the aluminum powder in the PCM.

As the main component of a phase change heat storage system, the phase change unit is a key part of the heat transfer efficiency of the system, and the heat transfer performance is closely related to its shape and structure. Common geometries for phase change heat storage units include square, cylindrical, spherical, and toroidal. The spherical shape has the largest heat transfer area for the same volume of heat storage unit. Ismail [16] visualized the solid–liquid interface inside a spherical capsule through experiments and numerical simulations. Asker [17] performed a numerical analysis of the inward solidification of PCM in spherical capsules. The results show that large-diameter spherical capsules have longer solidification times than small-diameter capsules. The entropy generation increases with increasing spherical capsule diameter and decreases after reaching a maximum.

The enhancement of heat transfer by adding fins to the heat storage unit has been investigated [18]. Li [19] installed fins on PCM shells in horizontal and vertical directions. The results showed that the triple fins in the vertical direction and double fins in the horizontal direction had the best melting enhancement for PCM with the best fire use efficiency. Koizumi [20] inserted copper plates inside spherical capsules to improve the latent heat storage rate of solid PCM and found that the latent heat storage rate was significantly increased. Bouguila [21] proposed a novel heat sink model consisting of nano-enhanced phase change materials and pin fins for the thermal management of electronic devices. The effects of fin thickness and fin number on the performance of PCM fins were investigated. The results showed that the increase in fin volume fraction and nanoparticle concentration did not improve the performance of the heat sink. Because of the high thermal conductivity of the fin, the dispersion of nanoparticles in the fin radiator had no significant effect on the thermal response. Fan [22] investigated the heat transfer process of PCM-confined melting in spherical capsules through experiments and numerical simulations and found a 30% reduction in melting time when the largest fin height was used. Aziz [23] used Ansys CFX to simulate the heat storage and release of PCM encapsulated in spheres, using pins and copper plating to improve the heat transfer of PCM within the spheres. The results showed that copper-plated spheres with pins had a 37% reduction in phase transition time compared to normal PCM spheres. Premnath [24] experimentally investigated frozen water in stainless-steel spherical containers with inner fins of different lengths. The results show that in a spherical vessel with a diameter of 75 mm, it is recommended to install four 13.5 mm–long heat sinks to achieve maximum heat flux.v. The highly conductive fins helped accelerate the phase change and obtain the best energy-saving effect. Meghari [25] investigated the melting process of PCM inside a spherical capsule using numerical simulation. The melting processes of regular fins and hollow fins at different angles were compared. The results showed that the fins and their geometry play a key role in the melting process. Changing the inclination angle of the fins does not always enhance the heat transfer. However, adding hollow fins is more effective than regular fins. Sharma [26] investigated the effect of fin position within a spherical capsule on melting time, with an enhancement of 107.28% when the fins were centrally located. Lou [27] studied the strengthening effects of metal foams and fins on ice storage spheres through numerical simulations. The temperature field, ice front evolution, solidification fraction, total solidification time, and cold storage capacity under four different strengthening modes (plane, fins, metal foam, and metal foam composite fins) were analyzed, and the optimal porosity was obtained. Amin [28] studied the enhanced heat transfer of a single PCM sphere in TES systems. The results showed that adding pins to the sphere reduced the charging time by 34%. Sun [29] proposed a new double-layer spherical phase change heat storage structural unit and investigated the effect of fins on its melting behavior by numerical simulation to elucidate the mechanism of enhanced heat transfer coupling between dual local natural convection and heat conduction in the spherical heat storage unit. Combining factors such as complete melting time and entropy production, the optimal installation position of the fins was determined as the outer side of the inner sphere. The optimal combination of double-layer spherical fins was obtained through single-factor analysis of the number, height, and thickness of double-layer spherical fins. For a spherical capsule with an outer diameter of 50 mm and an inner diameter of 30 mm, the optimal combination of fin parameters was Num = 3, H/L = 80%, and D = 0.5 mm.

Based on the above studies, it can be seen that there are various methods to enhance thermal storage technology. There are studies conducted for phase change materials, and there are also studies on the strengthening of thermal storage structures. However, mostly only one of the enhanced methods has been analyzed, with a lack of research on the combination of multiple factors. Doss [30] investigated the effect of fin length, bath temperature, and frozen mass on solidification time using the RSM. The results of the study are instructive for this paper. This study focuses on the addition of pin fins inside a spherical heat storage unit filled with nanocomposite phase change materials and uses the response surface method to comprehensively analyze the effect of different means on the enhanced heat transfer effect.

The objectives of this study were to (a) investigate the effect of pin fins on the melting characteristics of PCM in spherical thermal storage units; (b) determine the parameters affecting the melting time of PCM; and (c) develop a regression model for the melting time of PCM and predict new data after experimental confirmation.

2. Experimental Procedure

2.1. Design of Experiments

Paraffin mainly comprises straight-chain compounds. The latent heat and temperature of the phase transition of paraffin change with the growth of the carbon chain and increase with the growth of the carbon chain under normal circumstances. As the carbon chain grows and the melting point increases, the growth rate starts out fast, but it becomes slower and slower as the chain increases. However, the thermal conductivity of paraffin wax is low. Obviously, this lower thermal conductivity results in a low heat storage/release rate. In order to solve the problems of the low thermal conductivity, uneven temperature distribution, and low heat storage/release rate of pure paraffin PCMs, this study used composite PCMs to improve the thermal performance of paraffin PCMs.

Paraffin wax was used as the phase change material. With the "two-step" method, a certain amount of paraffin wax was weighed into a beaker and then placed in a constant temperature water bath at 80 °C to melt the paraffin wax completely, while a certain amount of CuO nanoparticles were weighed and added to the melted paraffin wax. The CuO nanoparticles were added to the melted paraffin wax and stirred with a magnetic stirrer for 30 min to make them fully integrated and then treated with an ultrasonic shaker for 60 min to disperse the CuO nanoparticles evenly in the paraffin wax to prepare a CuO/paraffin wax composite phase change material. Figure 1 shows the variation curve of the latent heat of phase transition of CuO/paraffin wax composite phase change materials in the case of different mass fractions measured using DSC.

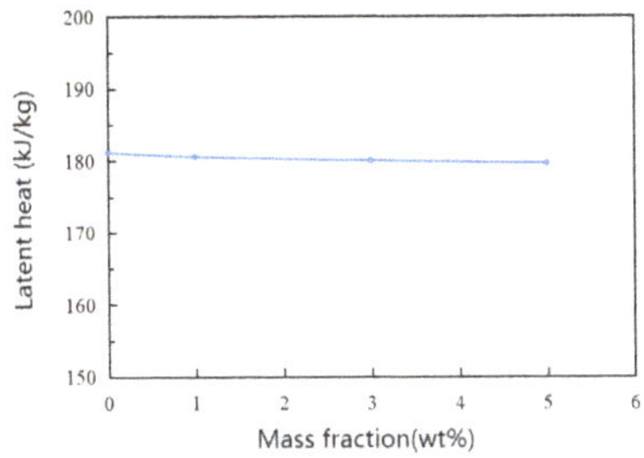

Figure 1. Curve of latent heat of phase change materials using DSC.

Two 304 stainless-steel hemispherical vessels were selected and flanged externally, and the two hemispheres were encapsulated into a complete spherical vessel using bolts, resulting in the spherical shell of the phase change heat storage unit used in this study. The inner diameter of the sphere was 97 mm, the thickness was 1.5 mm, and the sphere was welded with 3 mm–diameter pin fins made of aluminum alloy in quantities of 2, 4, and 6, as shown in Figure 2.

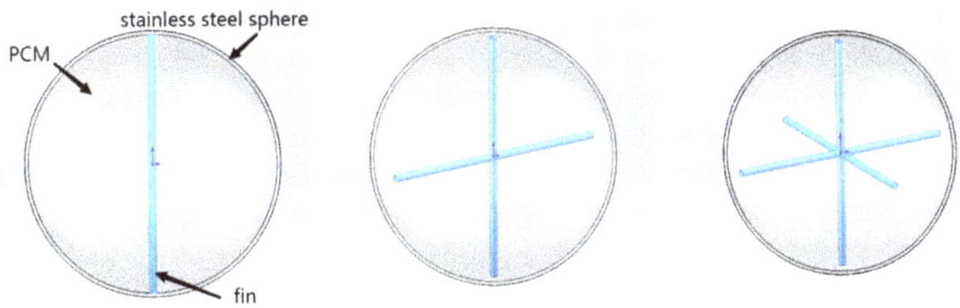

Figure 2. Distribution of fins inside the phase change heat storage sphere.

The composite phase change materials filled the heat storage sphere through small holes in the top of the sphere shell. The sphere shell was heated using an electric heating furnace to completely melt the composite phase change materials, which were added to the sphere in small quantities. Considering the thermal expansion of the phase change materials, the filling was terminated when the liquid phase change materials filled in the spheres reached a fixed height to ensure that no overflow would occur.

The experimental test system for melting within a spherical heat storage unit is shown in Figure 3. It mainly consisted of a constant temperature water bath, TP700 multi-channel data logger, thermocouples, and a heat storage sphere. K-type thermocouples were used for temperature testing and were arranged inside the sphere to record the internal temperature changes of the phase change materials during the melting process in real time. The locations of the thermocouples are shown in Figure 3. The duration the temperature at all monitoring points took to reach a steady-state value of 0.99 was taken as the melting time.

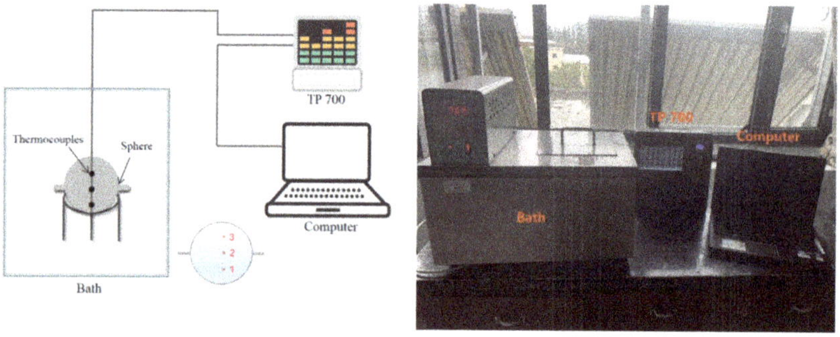

Figure 3. Diagram of melting experiment system.

At a water bath temperature of 70 °C, the temperature variation of pure paraffin with three fin quantities is shown in Figure 4a. Figure 4b shows the temperature variation in the heat storage unit for different mass fractions of CuO. For all cases, at the beginning of melting, the heat in the water bath was transferred to the solid PCM, which absorbed heat and warmed up. When the temperature approached the phase change temperature, the solid PCM started to undergo phase change, and all the absorbed heat was used for latent heat. Thus, the temperature underwent a period of slower change. When the solid PCM phase change was completely transformed into liquid PCM, the absorbed heat made the liquid PCM temperature rise. At the same time, owing to the influence of natural convection, the hotter liquid PCM moved to the upper part of the sphere so that the temperature of measurement point 3 rapidly rose to the water bath temperature and tended to stabilize. As the number of fins/CuO mass fraction increased, the heat transfer to measurement point 3 was faster, and the final time to reach the water bath temperature was shorter.

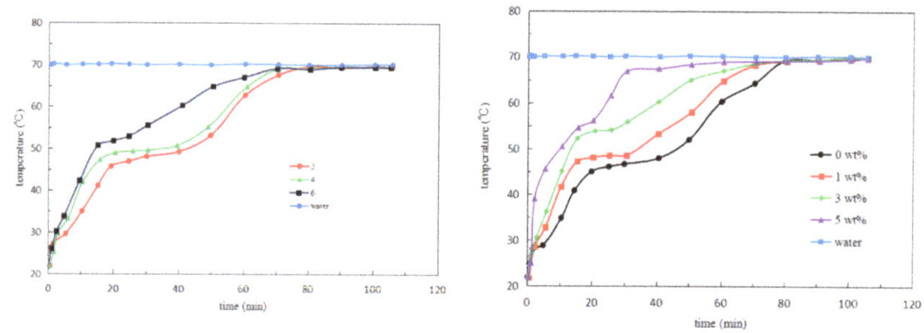

(**a**) (**b**)

Figure 4. Temperature change at measurement point 3 in the heat storage unit: (**a**) various numbers of fins; (**b**) different mass fractions of CuO.

2.2. Design of RSM

Response surface methodology (RSM) has been widely used in recent years for experimental design and optimization. RSM is an optimization method for experimental design that combines mathematics and statistics to enable correlation between the independent and dependent variables. The optimal working conditions are determined by analyzing the mathematical model established using RSM. Central composite design (CCD) is one of the most widely used experimental design methods in RSM that helps reduce the number of experiments used to achieve optimal conditions and analyze the interactions between parameters. In this study, CCD was applied to optimize the design of the complete melting time of phase change materials. The number of pin fins (N), water bath temperature (T), and the content of CuO in the nanocomposite phase change materials (M) were used as the factors for the study, and the effect of the three parameters on the complete melting time was analyzed using the response surface methodology. According to the central composite design principle, the values of three parameters were designed: $1 \le N \le 7$, $62\ °C \le T \le 78\ °C$, and 0 wt% $\le M \le 6$ wt%. The real values of each parameter were code-transformed to facilitate response surface analysis. After the coding conversion to determine the range of values, specific experimental conditions were designed, and the experiments were arranged on the melting test system according to the designed experimental conditions. The experimental data under each experimental condition were recorded. The coding levels of the influence factors of the operational parameters are shown in Table 1, and Table 2 shows the design scheme and results. For the measurement error of the measurement time, the experiment was conducted three times for each group under the same experimental conditions, and the melting time in Table 2 is the mean value of the three experiments. Table 3 shows the uncertainty analysis that existed during this experimentation

Table 1. Actual and coded values of the independent variables used for the experimental design.

Variable	Symbol	Real Values of Code Levels				
		$-\alpha$	−1	0	1	$-\alpha$
N	X_1	1	2	4	6	7
T (°C)	X_2	62	65	70	75	78
M (wt%)	X_3	0	1	3	5	6

Table 2. CCD experimental design and the obtained responses.

Run Number	Design Factors						Responses
	X_1		X_2		X_3		t (min)
	N		T (°C)		M (wt%)		
No.	Actual	Code	Actual	Code	Actual	Code	
1	6	+1	65	−1	1	−1	101.0
2	6	+1	65	−1	5	+1	97.3
3	6	+1	75	+1	1	−1	86.9
4	6	+1	75	+1	5	+1	78.9
5	2	−1	65	−1	1	−1	110.8
6	2	−1	65	−1	5	+1	106.9
7	2	−1	75	+1	1	−1	99.7
8	2	−1	75	+1	5	+1	93.6
9	7	+1.68	70	0	3	0	86.7
10	1	−1.68	70	0	3	0	99.8
11	4	0	78	+1.68	3	0	87.9
12	4	0	62	−1.68	3	0	105.8
13	4	0	70	0	6	+1.68	99.7
14	4	0	70	0	0	−1.68	96.8
15	4	0	70	0	3	0	94.4
16	4	0	70	0	3	0	94.8
17	4	0	70	0	3	0	95.6
18	4	0	70	0	3	0	93.5
19	4	0	70	0	3	0	93.0
20	4	0	70	0	3	0	94.6

Table 3. Uncertainty analysis.

Quantities	Uncertainties
Diameter	±0.02 mm
Temperature	±0.5 °C
Mass	±0.001 g

3. Results and Discussion

The experimental CCD included 20 groups. Groups 1–8 were orthogonal designs, and groups 9–14 formed a central composite design with $\alpha = \pm 1.68$. Groups 15–20 were centroid designs used to estimate the experimental error. The experimental data obtained from Table 2 were analyzed using the second-order polynomial model obtained from Equation (1).

$$y = \beta_0 + \sum_{i=1}^{3} \beta_i X_i + \sum_{i=1}^{3} \beta_{ii} X_i^2 + \sum_{j} \sum_{i=1\, i<j}^{3} \beta_{ij} X_i X_j + \varepsilon \quad (1)$$

where y is the predicted response value; X_i and X_j are coded independent variables; β_0, $\beta_1, \ldots, \beta_k$, β_{ij} are regression coefficients; and ε is the statistical error. The least-squares method was used to determine the regression coefficients β. For each response, regression coefficients were obtained using coded variables, and then regression coefficients were calculated for the actual variables. Analysis of variance (ANOVA) was used to determine the valid parameters of the model and to interpret its significance.

After the regression fitting of each operational parameter, the regression model for the coding level of the complete melting time t is obtained as

$$t = 94.04 - 5.32X_1 - 6.52X_2 - 2.53X_3 - 1.01X_{12} - 0.2125X_{13} - 0.8125X_{23} + 0.2274{X_1}^2 + 1.54{X_2}^2 + 0.3385{X_3}^2 \quad (2)$$

The regression model for the actual level of complete melting time is

$$t = 458.09 + 4.13N - 9.30T + 4.13M - 0.101NT - 0.053NM - 0.081TM + 0.057N^2 + 0.062T^2 + 0.085M^2 \quad (3)$$

From Equation (2), it can be seen that the water bath temperature has a greater effect on the melting time because the absolute value coefficient of the water bath temperature is greater than the absolute value coefficient of the number of pin fins and the mass fraction of CuO nanoparticles. Although the three primary term coefficients are negative, the required melting time is as short as possible, so all three factors have a coordinating effect on the complete melting time. In the actual level of the quadratic regression mathematical model, comparing the magnitude of each interaction term coefficient, the order of interaction between interaction term factors can be derived from Equation (3), which shows that the order of interaction between interaction factors is $M^2 > T^2 > N^2 > NT > NM > TM$.

Table 4 shows the ANOVA data for the complete melting time. The F-value was used to determine the statistical significance of the second-order regression model. When $p < 0.05$, the corresponding parameters were all significant, and when $p < 0.0001$ and F = 34.37 in the complete melting time model, the model was apparently significant. The coefficient of determination R^2 is the ratio of the response contribution of the regression model, and the closer it is to one, the higher the model explanation. $R^2 = 0.9687$. This indicates that more than 96.87% of the data deviations can be explained. The adequacy of the model can be tested by the correction decision coefficient R^2_{adj}, and $R^2_{adj} = 0.9405$, indicating a strong interaction between the predicted response values and the experimental results. AP is defined as a measure of the ratio of signal to noise, and AP > 4 is desirable. The model has AP = 22.2023, indicating that the regression equation is well-adapted.

Table 4. Results of ANOVA.

Source	Sum of Square	Degree of Freedom	Mean Square	F-Value	p-Value	R^2	Adjusted R^2	Adequate Precision
Model	1036.96	9	115.22	34.37	<0.0001	0.9687	0.9405	22.2023
Error	4.37	5	0.8737	-	-	-	-	-
Total	1070.48	19	-	-	-	-	-	-

The residuals are the values fitted by the regression model subtracted from the experimental measurements. The smaller the residuals, the more accurately the regression model describes the experimental results. Figure 5a shows the relationship between the experimental values and the response prediction values. From the ideal linear distribution curve, we can see that the experimental and simulated values are basically distributed along a straight line, indicating that the experimental and simulated values are highly fitted. A normal plot of the experimental residuals can be used to determine the fit of the model, as shown in Figure 5b, which shows the fit of the residuals to the standard deviation. If the distribution of points on the plot is a straight line, the residuals are normally distributed. As can be seen from the figure, most of the experimental values fall on the predicted values in a linear distribution, indicating that the model fits the actual results well. Figure 5c,d shows the relationship between the residuals and the predicted values. The points in the diagnostic plot that are beyond the upper and lower boundary lines are anomalies, and these anomalies can indicate whether the model has large deviations. It can be seen from the figure that no points are located outside the boundary line, indicating that the model and the data match. Moreover, there are no large interference points in the model points, and the distribution of model points is also concentrated, so the model diagnostic plot results match the ANOVA data.

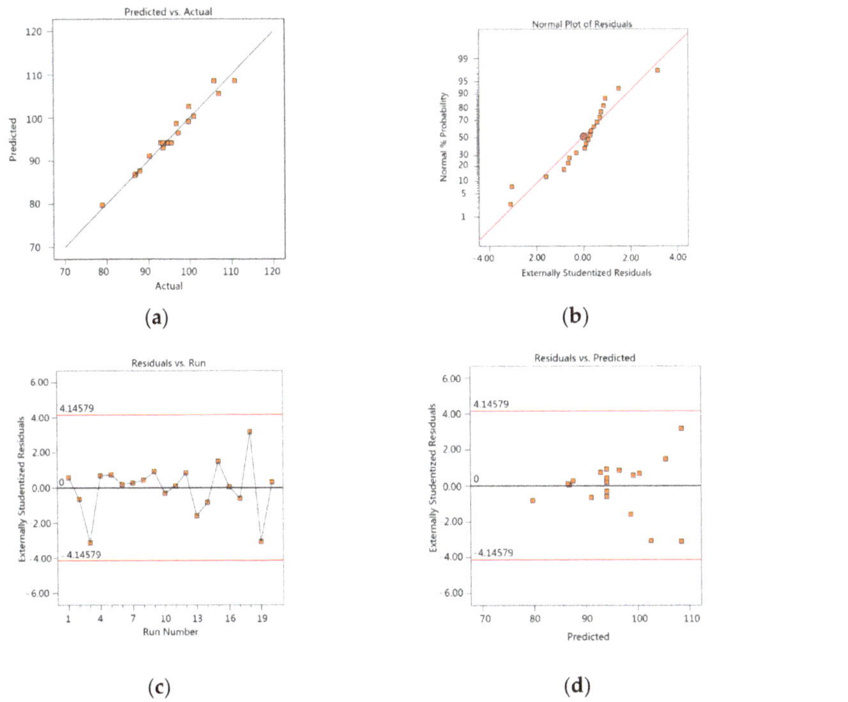

Figure 5. The diagnostic plots of RSM models: (**a**) actual vs. predicted values for complete melting time; (**b**) normal probability plots; (**c**) residuals vs. experimental runs; (**d**) residuals vs. predicted values.

Response surfaces are used to analyze the interaction between the factors and the optimal response values at the optimal factor conditions. Using Design-Expert 13.0, 3D surface plots of the response values were obtained, and the contour lines were circular for insignificant interaction of the two factors and elliptical for significant interaction. The melting time of PCM in a spherical thermal storage unit is affected by the number of pin fins, the water bath temperature, and the content of CuO nanoparticles, all three of which have a synergistic effect on the complete melting time.

Figure 6 shows the effect of the number of pin fins and the water bath temperature on the melting time. The complete melting time decreases with increasing the number of pin fins and increases with decreasing the water bath temperature, and the water bath temperature has a greater effect on the melting time than the number of pin fins. This is because the melting of PCM is mainly influenced by temperature; the higher the external ambient temperature, the faster the PCM reaches the phase change point. Therefore, if the water bath temperature is high, the heat that can be absorbed by the PCM in the spherical unit increases; the melting rate is faster, and the melting time decreases. During the melting of the phase change material, the melting of the material in the units is faster because of the large temperature difference between the inner wall of the sphere and the solid phase change material, and the heat transfer is mainly heat conduction. As melting proceeds, the thickness of the liquid PCM layer increases, leading to an increase in thermal resistance and resulting in a gradual decrease in the melting rate. At this point, the addition of internal pin ribs is equivalent to increasing the heat transfer area in contact with the inside of the sphere, which speeds up the melting process to a certain extent and shortens the melting time.

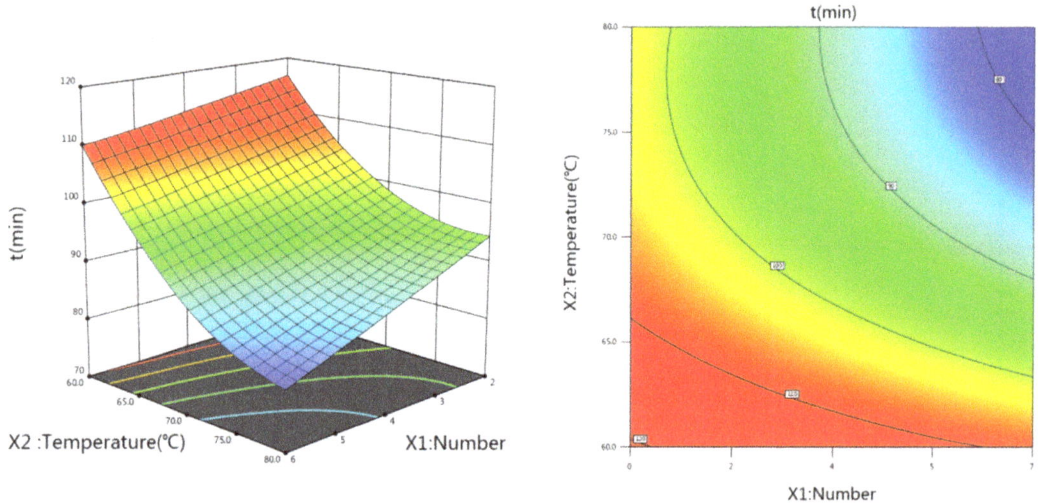

(a) (b)

Figure 6. Response surface (**a**) and contour plot (**b**) of the number of pin fins and water bath temperature on the complete melting time.

Figure 7 shows the interaction effect of the number of pin fins and the mass fraction of CuO nanoparticles, which is not significant, as shown by the circular shape of the contour plot. From Equation (2), it is clear that the number of pin fins has a greater effect on the melting time of PCM than the CuO nanoparticles content. The melting process of PCM within an externally heated spherical unit is quite complex. The unmelted solid moves because of the density difference between the solid and liquid phases, and typically, contact melting occurs at the bottom of the spherical unit because of the sinking of the heavier solid PCM. At this time, adding pin fins inside the spherical unit speeds up the melting of the internal PCM, and the more pin fins, the stronger the internal heat transfer effect, which correspondingly shortens the melting time of the PCM.

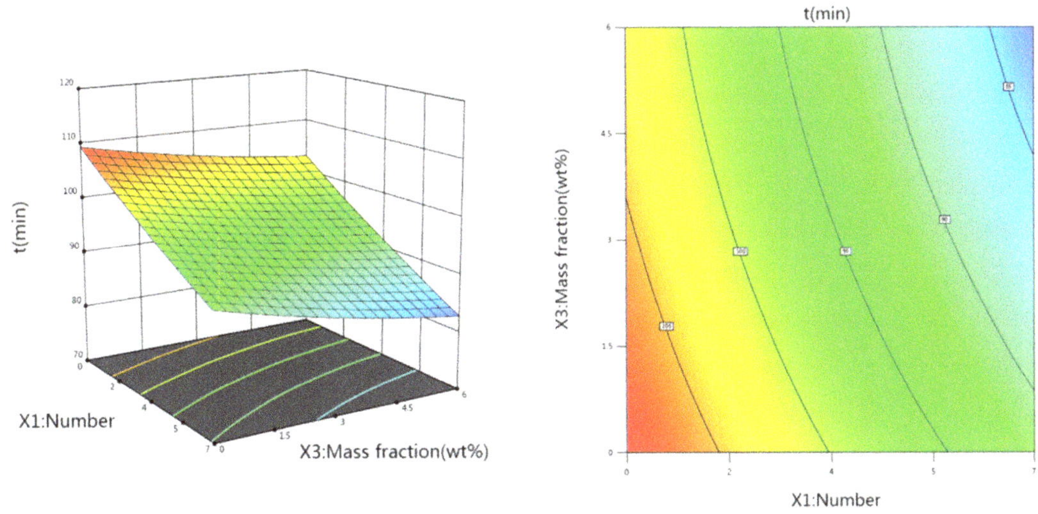

(**a**) (**b**)

Figure 7. Response surface (**a**) and contour plot (**b**) of the number of pin fins and mass fraction of CuO nanoparticles on the complete melting time.

The increase in water bath temperature and the increase of CuO nanoparticles content in Figure 8 both accelerate the melting of PCM within the spherical thermal storage unit accordingly. The addition of CuO nanoparticles increases the thermal conductivity of PCM and enhances the melting heat transfer within the spherical thermal storage unit. However, the change in thermal conductivity has a relatively small effect on the melting time compared to the increase in water bath temperature.

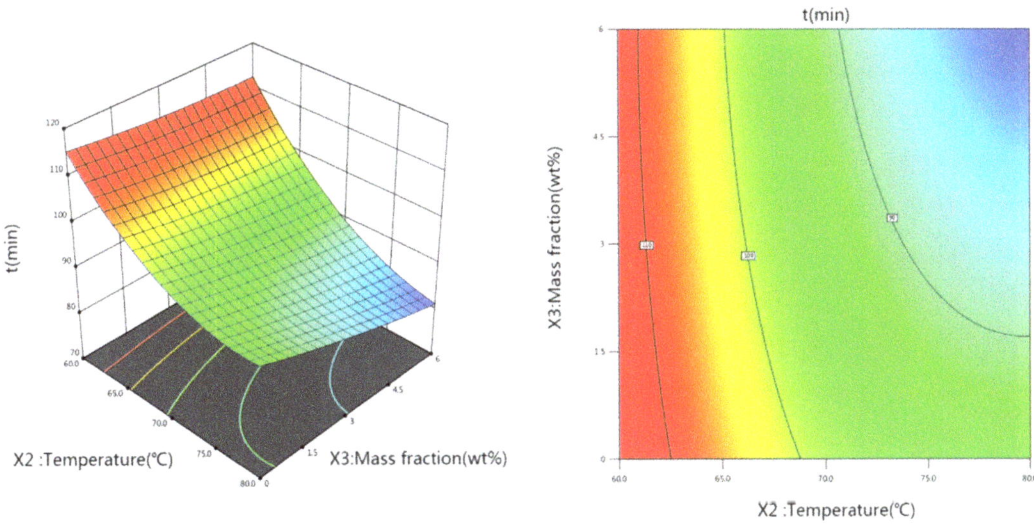

(**a**) (**b**)

Figure 8. Response surface (**a**) and contour plot (**b**) of water bath temperature and CuO nanoparticles mass fraction on the complete melting time.

The main objective of optimization using the response surface method is to obtain the shortest melting time. Within the range of values of the three factors, the optimum combination is chosen such that the value of melting time t is minimized. Optimization analysis of the system was carried out using Design-Expert software to obtain the optimum level of influence factors of the system as well as the optimum melting time t.

The predicted value of t was 79.7 min, and the optimal level of each influence factor was obtained from the optimization analysis: $N = 6$, $T = 75$ °C, and $M = 5$ wt%. In this study, response surface methodology was applied to simulate and optimize the operational impact parameters of the melting process in a spherical thermal storage unit with a minimum number of experiments, and the mathematical analysis of variance of the response surface model verified the accuracy of the model. Experiments showed that the response surface method is accurate and reliable for optimizing the operational parameters of PCM melting in spherical thermal storage units.

4. Conclusions

In this study, by building an experimental system for melting spherical heat storage units, selecting the number of pin fins, water bath temperature, and the content of CuO nanoparticles added in paraffin as variables and the melting time of composite PCM as the target value, the following conclusions were obtained using the response surface methodology to optimize the target value:

(1) CCD was used to design the experimental conditions, and the data were imported into Design-Expert software to obtain the quadratic polynomial regression model of melting time. The analysis of variance showed that the regression model was significant, and the comparison analysis between the experimental and predicted values of melting time further proved that the obtained model had high reliability in fitting the melting time throughout the experimental design range.

(2) The effect of the interaction between the factors on the melting time was analyzed using the response surface methodology. Among them, the water bath temperature has a large effect on the response value of PCM melting time in the spherical heat storage unit, and the interaction between the water bath temperature and the number of pin fins has the most significant effect on the melting time. The addition of CuO nanoparticles enhances heat transfer, but the effect is not particularly significant compared to the other two influencing parameters.

(3) Using the optimal module of Design-Expert software, the optimal operating conditions of the system were obtained as follows: six pin fins were added to the spherical thermal storage unit, the water bath temperature was 75 °C, the mass fraction of CuO nanoparticles was 5 wt%, and the melting time of PCM was the shortest, 79.7 min. Using the response surface method, a model for predicting the melting time was further developed, and the results of this study are useful for the design of the spherical thermal storage unit. The results of this study will have a guiding significance for the design of spherical thermal storage units.

Computational analysis using the RSM is used to improve and optimize the statistical properties of the optimization objective to some extent. This research has achieved some results; however, there is still some distance from the practical application. The spherical phase change heat storage unit is not used alone in practical applications, and subsequent research should continue to explore other influencing factors that may affect the heat storage and release performance of the spherical phase change heat storage unit for in-depth study.

Author Contributions: Conceptualization, R.T.; methodology, L.L.; software, L.L.; validation, X.G.; formal analysis, R.T.; investigation, Y.Z.; resources, R.T.; data curation, L.L.; writing—original draft preparation, L.L.; writing—review and editing, R.T.; visualization, X.G.; supervision, Y.Z.; project administration, L.L.; funding acquisition, X.G. All authors have read and agreed to the published version of the manuscript.

Funding: This research was funded by the Inner Mongolia Autonomous Region Natural Science Fund, grant number 2023QN05039 and the Scientific Research Projection of Higher Schools of Inner Mongolia, grant number NJZY23115.

Institutional Review Board Statement: Not applicable.

Informed Consent Statement: Not applicable.

Data Availability Statement: The data presented in this study are available on request from the corresponding author.

Conflicts of Interest: The authors declare no conflict of interest.

References

1. Du, K.; Calautit, J.; Wang, Z.; Wu, Y.; Liu, H. A review of the applications of phase change materials in cooling, heating and power generation in different temperature ranges. *Appl. Energy* **2018**, *220*, 242–273. [CrossRef]
2. Jesumathy, S.; Udayakumar, M.; Suresh, S. Experimental study of enhanced heat transfer by addition of CuO nanoparticle. *Heat Mass Transf.* **2012**, *48*, 965–978. [CrossRef]
3. Wang, J.; Li, Y.X.; Wang, Y.; Yang, L.; Kong, X.F.; Sunden, B. Experimental investigation of heat transfer performance of a heat pipe combined with thermal energy storage materials of CuO-paraffin nanocomposites. *Solar Energy* **2020**, *211*, 928–937. [CrossRef]
4. Nourani, M.; Hamdami, N.; Keramat, J.; Moheb, A.; Shahedi, M. Thermal behavior of paraffin-nano-Al_2O_3 stabilized by sodium stearoyl lactylate as a stable phase change material with high thermal conductivity. *Renew. Energy* **2016**, *88*, 474–482. [CrossRef]
5. Yadav, C.; Sahoo, R.R. Thermal analysis comparison of nano-additive PCM-based engine waste heat recovery thermal storage systems: An experimental study. *J. Therm. Anal. Calorim.* **2021**, *147*, 2785–2802. [CrossRef]
6. Sahan, N.; Paksoy, H.O. Thermal enhancement of paraffin as a phase change material with nanomagnetite. *Sol. Energy Mater. Sol. Cells* **2014**, *126*, 56–61. [CrossRef]
7. Hu, Y.W.; Shi, L.; Zhang, Z.D.; He, Y.R.; Zhu, J.Q. Magnetic regulating the phase change process of Fe_3O_4-paraffin wax nanocomposites in a square cavity. *Energy Convers. Manag.* **2020**, *213*, 112829. [CrossRef]
8. Zhang, Y.; Zheng, S.; Zhu, S.; Ma, J.; Sun, Z.; Farid, M. Evaluation of paraffin infiltrated in various porous silica matrices as shape-stabilized phase change materials for thermal energy storage. *Energy Convers. Manag.* **2018**, *171*, 361–370. [CrossRef]
9. Kumar, P.M.; Mylsamy, K.; Saravanakumar, P.T. Experimental investigations on thermal properties of nano-SiO2/paraffin phase change material (PCM) for solar thermal energy storage applications. *Energy Source Part A-Recovery Util. Environ.-Ronmental Eff.* **2019**, *42*, 2420–2433.
10. Tian, B.; Yang, W.; Luo, L.; Wang, J.; Zhang, K.; Fan, J.; Wu, J.; Xing, T. Synergistic enhancement of thermal conductivity for expanded graphite and carbon fiber in paraffin/EVA form-stable phase change materials. *Sol. Energy* **2016**, *127*, 48–55. [CrossRef]
11. Chen, M.; He, Y.; Ye, Q.; Zhang, Z.; Hu, Y. Solar thermal conversion and thermal energy storage of CuO/Paraffin phase change composites. *Int. J. Heat Mass Transf.* **2019**, *130*, 1133–1140. [CrossRef]
12. Lin, S.C.; Al-Kayiem, H.H. Evaluation of copper nanoparticles—Paraffin wax compositions for solar thermal energy storage. *Sol. Energy* **2016**, *132*, 267–278. [CrossRef]
13. Kumar, P.M.; Mylsamy, K.; Prakash, K.B.; Nithish, M.; Anandkumar, R. Investigating thermal properties of Nanoparticle Dispersed Paraffin(NDP)as phase change material for thermal energy storage. *Mater. Today Proc.* **2020**, *45*, 745–750. [CrossRef]
14. Pasupathi, M.K.; Alagar, K.; Mm, M.; Aritra, G. Characterization of Hybrid-nano/Paraffin Organic Phase Change Material for Thermal Energy Storage Applications in Solar Thermal Systems. *Energies* **2020**, *13*, 5079. [CrossRef]
15. Li, W.; Wang, Y.-H.; Kong, C.-C. Experimental study on melting/solidification and thermal conductivity enhancement of phase change material inside a sphere. *Int. Commun. Heat Mass Transf.* **2016**, *68*, 276–282. [CrossRef]
16. Ismail, M.; Alkhazaleh, A.H.; Masri, J.; Ali, A.M.; Ali, M. Experimental and numerical analysis of paraffin waxes during solidifi-cation inside spherical capsules. *Therm. Sci. Eng. Prog.* **2021**, *26*, 101095. [CrossRef]
17. Asker, M.; Ganjehsarabi, H.; Coban, M.T. Numerical investigation of inward solidification inside spherical capsule by using temperature transforming method. *Ain Shams Eng. J.* **2018**, *9*, 537–547. [CrossRef]
18. Fan, L.; Khodadadi, J.M. Thermal conductivity enhancement of phase change materials for thermal energy storage: A review. *Renew. Sustain. Energy Rev.* **2011**, *15*, 24–46. [CrossRef]
19. Li, J.; Abdulghani, Z.R.; Alghamdi, M.N.; Sharma, K.; Niyas, H.; Moria, H.; Arsalanloo, A. Effect of twisted fins on the melting performance of PCM in a latent heat thermal energy storage system in vertical and horizontal orientations: Energy and exergy analysis. *Appl. Therm. Eng.* **2023**, *219*, 119489. [CrossRef]
20. Koizumi, H. Time and spatial heat transfer performance around an isothermally heated sphere placed in a uniform, down-wardly directed flow (in relation to the enhancement of latent heat storage rate in a spherical capsule). *Appl. Therm. Eng.* **2004**, *24*, 2583–2600. [CrossRef]
21. Bouguila, M.; Samet, A.; Ben Souf, M.A.; El Hami, A.; Haddar, M. Thermal performances of finned heat sink filled with nano-enhanced phase change materials: Design optimization and parametric study. *Int. J. Heat Mass Transf.* **2023**, *202*, 123710. [CrossRef]

22. Fan, L.-W.; Zhu, Z.-Q.; Xiao, S.-L.; Liu, M.-J.; Lu, H.; Zeng, Y.; Yu, Z.-T.; Cen, K.-F. An experimental and numerical investigation of constrained melting heat transfer of a phase change material in a circumferentially finned spherical capsule for thermal energy storage. *Appl. Therm. Eng.* **2016**, *100*, 1063–1075. [CrossRef]
23. Aziz, S.; Amin, N.; Majid, M.A.; Belusko, M.; Bruno, F. CFD simulation of a TES tank comprising a PCM encapsulated in sphere with heat transfer enhancement. *Appl. Therm. Eng.* **2018**, *143*, 1085–1092. [CrossRef]
24. Premnath, D.; Chandrasekaran, P.; Subramanian, L.R.G.; Senthil, R. Experimental performance of a finned spherical container in cold thermal storage for tropical buildings. *Environ. Sci. Pollut. Res.* **2022**, *29*, 76793–76804. [CrossRef]
25. Meghari, Z.; Bouhal, T.; Benghoulam, M.; El Rhafiki, T.; El Khattabi, E.M.; Doghmi, H.; Mohammed, O.J. Numerical simulation of a phase change material in a spherical capsule with a hollow fin. *J. Energy Storage* **2021**, *43*, 103024. [CrossRef]
26. Sharma, A.; Kothari, R.; Sahu, S.K. Effect of fin location on constrained melting heat transfer of phase change material in a spherical capsule: A numerical study. *J. Energy Storage* **2022**, *52*, 104922. [CrossRef]
27. Lou, X.; Wang, H.; Xiang, H. Solidification performance enhancement of encapsulated ice storage system by fins and copper foam. *Int. J. Refrig.* **2022**, *134*, 293–303. [CrossRef]
28. Amin, N.A.M.; Bruno, F.; Belusko, M. Investigation of Conducting Pins in Sphere Filled with Phase Change Material for Enhancing Heat Transfer in Thermal Energy Storage. *Adv. Mater. Res.* **2012**, *472–475*, 1693–1697. [CrossRef]
29. Sun, B.; Sun, M.; Gao, L.; Che, D.; Deng, Y. Thermal performance analysis and optimization of a double-layer spherical phase-change material capsule with annular fins. *J. Energy Storage* **2023**, *57*, 106280. [CrossRef]
30. Doss, P.; Ponnusamy, C.; Ramachandran, G.S.L. Predictive modeling of solidification characteristics of a phase change material in a metallic spherical capsule fitted with fins of different lengths. *Energy Sources Part A Recover. Util. Environ. Eff.* **2020**, 1–14. [CrossRef]

applied sciences

MDPI

Article

Research and Application of Steam Condensation Heat Transfer Model Containing Noncondensable Gas on a Wall Surface

He Li [1], Xiaoming Yang [1,*], Chen Wang [1], Shang Shi [2], Rubing Ma [1] and Yidan Yuan [1]

1 CNNC Key Laboratory on Severe Accident in Nuclear Power Safety, China Nuclear Power Engineering Co., Ltd., No 117 Xisanhuan North Road, Beijing 100084, China; lihea@cnpe.cc (H.L.); wangchen@cnpe.cc (C.W.); marb@cnpe.cc (R.M.); yuanyd@cnpe.cc (Y.Y.)

2 State Key Laboratory of Multiphase Flow in Power Engineering, Xi'an Jiaotong University, Xian-Ning West Road 28#, Xi'an 710049, China; 18700751635@stu.xjtu.edu.cn

* Correspondence: yangxm@cnpe.cc; Tel.: +86-010-88023687

Abstract: Steam condensation plays an important role in various engineering processes due to its excellent heat transfer performance. However, condensation in the presence of noncondensable gas has attracted great attention in recent years since noncondensable gas will have a negative effect on condensation heat transfer. The present study proposes a comprehensive model coupled with convective heat transfer, liquid film heat transfer and steam condensation for the heat transfer of condensation with noncondensable gas and uses it in the Program Integrated for Severe Accident Analysis (PISAA) for a nuclear power plant. The condensation heat transfer model has good universality, the calculation process is stable with less iteration and a fast convergence and it is verified and validated by comparing the simulation results of the PISAA and those from traditional containment analysis codes, as well the experiments from the Wisconsin condensation tests; then, a sensitivity analysis for the parameters of the heat transfer coefficient is performed. The validation results show that the average error of the condensation heat transfer coefficient is approximately 10%, and the maximum error does not exceed 30%. The deviation from the experimental data is limited in the acceptable range, which could fulfill the requirement for the analysis of containment accidents in nuclear power plants.

Keywords: steam condensation; two-phase flow; noncondensable gas; heat and mass transfer; heat structure

Citation: Li, H.; Yang, X.; Wang, C.; Shi, S.; Ma, R.; Yuan, Y. Research and Application of Steam Condensation Heat Transfer Model Containing Noncondensable Gas on a Wall Surface. *Appl. Sci.* **2023**, *13*, 10520. https://doi.org/10.3390/app131810520

Academic Editor: Francesca Scargiali

Received: 16 August 2023
Revised: 15 September 2023
Accepted: 19 September 2023
Published: 21 September 2023

1. Introduction

Steam condensation heat transfer is a common phenomenon in daily life and is widely used in the chemical industry, refrigeration, energy and other engineering processes. However, steam is always mixed with some gases that may not condense under different working conditions, while noncondensable gas can have a negative effect on condensation heat transfer [1–4].

The presence of noncondensable gases (such as air) in steam, even at very low levels, can significantly worsen heat transfer in applications [4]. Taking nuclear power plant accidents as an example, the release of high-temperature and high-pressure gas from the primary system will result in a sudden increase in temperature and pressure in containment, thereby endangering the structural integrity of the containment. The condensation heat transfer of high-temperature and high-pressure steam on the containment wall is an important method to remove internal heat; however, the presence of any noncondensable gas, such as air, can greatly affect the steam condensation efficiency, lead to a deterioration of heat transfer performance and complicate the overall process [2,3]. Consequently, it is of great significance to analyze the factors affecting condensation so as to improve the calculation accuracy of condensation heat transfer.

The condensation process of steam mixed with noncondensable gas is very complicated, as it involves not only the exchange and transfer of mass, energy and momentum between steam and condensed liquid, but also the exchange and diffusion of energy and momentum between steam and noncondensable gas [1]. So far, the research on this process mainly consists of experimental studies [5–15] and theoretical analyses [16–28].

In the experimental aspect, researchers have paid much attention to the heat transfer characteristics of condensation under various working conditions [6,7,10,14,15], as well as heat transfer enhancement using variant means. They conducted many experiments and obtained different heat transfer correlations, like the commonly used Uchida [7], Tagami [8] and Dehbi [9] models and so on, and correlations were made with the heat transfer rate via gas concentration, pressure, temperature, surface subcooling, and some other parameters. However, the main obstacle is that most correlations are not universally valid, and each empirical relation has a very strict application condition.

At the same time, a numerical calculation has become an important means to explain condensation phenomena by means of both lumped parameter codes and 3D CFD codes [20,21]; in contrast to the experiments, the numerical method can obtain more detailed information about the flow characteristics and gas concentration distribution. And many factors have also been considered to affect the condensation efficiency [25–28] like the geometrical parameter, thickness of film, suction factor and so on, but for pure theoretical numerical calculations, the calculation process is complicated and difficult to understand, and at the same time, too many iterations lead to a divergence of the calculation results.

In a nuclear power plant, the condensation process of steam with noncondensable gas on the surface of a heat structure is accompanied by an intense heat and mass transfer, which directly affects the spatial distribution of the temperature and pressure in containment, and then affects the safe operation of the integrated reactor [4]. This paper analyzes and discusses this phenomenon, proposes a condensation calculation model, and applies this calculation model to the integral analysis code of the PISAA (Program Integrated for Severe Accident Analysis) for severe accidents. By conducting specific working conditions, the wall condensation model in the PISAA is compared and verified with the mainstream containment thermal hydraulic codes. Additionally, the calculation results of the model are validated by comparing the data with those of the Wisconsin condensation experiment [15]. Finally, a brief sensitivity analysis is performed on the condensation heat transfer model, which further improves the accuracy of the condensation heat transfer calculation.

2. Condensation Heat Transfer Model

Figure 1 shows the condensation model of steam with noncondensable gas on the vertical wall. For the condensation process containing noncondensable gas, when the wall temperature is lower than the saturation temperature corresponding to the partial pressure of steam in the mixture, the saturated steam will condensate on the wall and form a liquid film. Meanwhile, the liquid film flows along the direction of gravity, and its flow state changes from laminar to turbulent as the liquid film thickness increases [1,14]. Due to the accumulation of noncondensable gas on the surface of the liquid film, when the mixture moves towards the interface, the aggregation of noncondensable gas molecules causes an increase in the gas's partial pressure and forms the driving force of the reverse diffusion of noncondensable gas to the mainstream gas. While the steam condenses and its partial pressure decreases below that of the steam in the mainstream gas at the interface between the phases, this pressure differential promotes the diffusion of steam towards the interface, which is facilitated by the pressure gradient. The diffusion of steam towards the condensing surface and the noncondensable gases towards the mainstream gas maintain a dynamic equilibrium with a constant total pressure [12]. Under the influence of two conditions, the state parameters at the phase interface cannot be determined.

The steam in the gas mixture is transferred to the wall by means of mass diffusion, which requires it to pass through the highly concentrated noncondensable gas layer that gathers on the surface of the condensate film, and then condenses on the surface and releases latent heat. The heat transfer resistance of the whole steam condensation is mainly divided into three parts, and the steam transfers the heat to the surface of the liquid film by means of condensation and convection, respectively, that is, there is condensation thermal resistance and convection thermal resistance, and further heat passes through the liquid film thermal resistance and finally reaches the condensing wall.

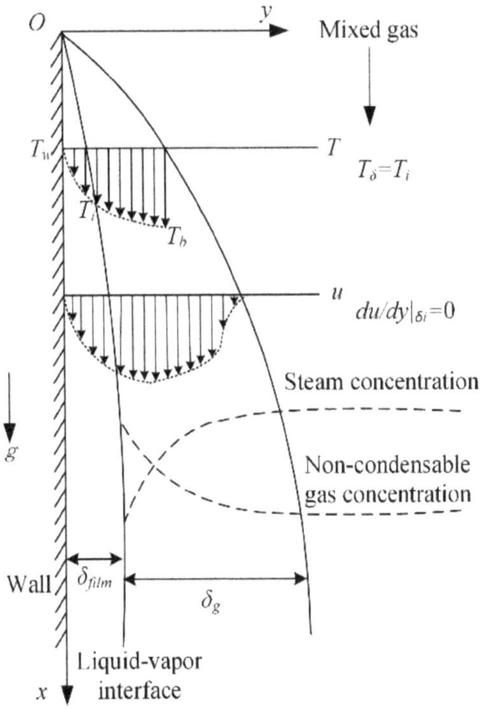

Figure 1. Condensation model of steam with noncondensable gas on vertical wall.

Correspondingly, the total heat transfer coefficient is determined by the thermal resistance of the three aspects above: the liquid film heat transfer coefficient h_{film} determined by $T_i - T_w$, the convective heat transfer coefficient h_{conv} and the condensation heat transfer coefficient h_{cond} determined by $T_g - T_i$. The energy conservation equation is as follows [29]:

$$\frac{q}{h_{\mathrm{tot}}} = \frac{q}{h_{\mathrm{film}}} + \frac{q}{h_{\mathrm{conv}} + h_{\mathrm{cond}}} \quad (1)$$

$$h_{\mathrm{tot}}(T_g - T_w) = h_{\mathrm{film}}(T_i - T_w) = (h_{\mathrm{conv}} + h_{\mathrm{cond}})(T_g - T_i) \quad (2)$$

where: q—total heat transfer flux;
h_{tot}—total heat transfer coefficient;
h_{film}—film heat transfer coefficient;
h_{conv}—convective heat transfer coefficient;
h_{cond}—steam condensation heat transfer coefficient;
T_g—mixture gas temperature;
T_i—interface temperature;
T_w—surface temperature of the plate.

As can be seen from Equation (2), the total heat transfer coefficient h_{tot} is related to the liquid film heat transfer coefficient h_{film}, convective heat transfer coefficient h_{conv} and steam condensation heat transfer coefficient h_{cond}, as well as the mixture gas temperature, interface temperature and surface temperature. So, in order to obtain the final total heat transfer coefficient, it is necessary to solve the h_{film}, h_{conv} and h_{cond} one by one, and determine the interface temperature T_{i}.

For the film heat transfer coefficient h_{film}, generally, the Nusselt theory and the modified equation based on the Nusselt theory are widely used to calculate the heat transfer coefficient of liquid film. However, according to the Nusselt theory, the liquid film is assumed to be laminar flow; when the vertical surface is too long, the liquid film fully develops, and the flow in the liquid film gradually changes from laminar flow to wavy laminar flow and then to turbulence flow. Using the solution based on the Nusselt theory will simply produce a certain deviation, so the following formula is adopted for laminar flow when $Re < 30$ [29]:

$$Re = 3.78\left[\frac{k_{\text{l}}L(T_{\text{i}} - T_{\text{w}})}{\mu_{\text{l}}h^{*}_{\text{fg}}(v_{\text{l}}^{2}/g)^{1/3}}\right]^{3/4} \quad Re < 30 \tag{3}$$

Kutateladze's relation is adopted for wavy laminar flow when $30 \leq Re < 1800$:

$$Re = \left[\frac{3.70k_{\text{l}}L(T_{\text{i}} - T_{\text{w}})}{\mu_{\text{l}}h^{*}_{\text{fg}}(v_{\text{l}}^{2}/g)^{1/3}} + 4.81\right]^{0.820} \quad 30 \leq Re < 1800 \tag{4}$$

Labuntsov's relation is adopted for turbulent condensate flow when $Re \geq 1800$:

$$Re = \left[\frac{0.069k_{\text{l}}L(T_{\text{i}} - T_{\text{w}})}{\mu_{\text{l}}h^{*}_{\text{fg}}(v_{\text{l}}^{2}/g)^{1/3}}Pr^{0.5} - 151Pr^{0.5} + 253\right]^{4/3} \quad Re \geq 1800 \tag{5}$$

where: k_{l}—thermal conductivity of the liquid;
L—height of the vertical plate;
μ_{l}—dynamic viscosity of the liquid;
v_{l}—kinematic viscosity of the liquid;
g—gravitational acceleration;
Pr—Prandtl number.

Actually, the condensation process is cooled further to some average temperature between T_{i} and T_{w}, releasing more heat in the process. Therefore, the actual heat transfer will be larger. Rohsenow suggested that the cooling of the liquid below the saturation temperature can be accounted for the modified latent heat of vaporization h^{*}_{fg}, defined as $h^{*}_{fg} = h_{\text{fg}} + 0.68c_{p,\text{l}}(T_{\text{i}} - T_{\text{w}})$, where $c_{p,\text{l}}$ is the specific heat of the liquid at the average film temperature.

After calculating Re, the liquid film heat transfer coefficient h_{film} is finally calculated as follows:

$$h_{\text{film}} = \frac{Re\mu_{\text{l}}h^{*}_{\text{fg}}}{4L(T_{\text{i}} - T_{\text{w}})} \tag{6}$$

For the convective heat transfer coefficient h_{conv}, according to the numerical correlation of the convective heat transfer similarity criterion without considering the convective heat transfer coefficient under the influence of normal mass transfer, it can be calculated as follows:

$$h_{\text{conv}} = Nu \cdot k/l \tag{7}$$

The Nu number in the equation above refers to the relationship between turbulent natural convection and forced convection over a flat plate according to different convection forms:

$$Nu^{FC} = 0.037Re^{0.8}Pr^{1/3} \tag{8}$$

$$Nu^{NC} = 0.13(Gr \cdot Pr)^{1/3} \tag{9}$$

For mixed convection, the following equation is adopted [29]:

$$Nu^{n}_{\text{mixed}} = Nu^{n}_{\text{FC}} \pm Nu^{n}_{\text{NC}} \tag{10}$$

where Nu_{mixed} is the Nusselt number for the mixed flow, and Nu_{FC} and Nu_{NC} are the Nusselt number calculated via forced convection correlation and natural convection correlation under given conditions, respectively. A positive sign was taken when the two flow directions were the same, a negative sign was taken when they were opposite, and for an uncertain flow direction, the minimum value of Nu^{n}_{mixed} was adopted via conservative estimation. The exponent, n, is usually 3.

For the steam condensation heat transfer coefficient h_{cond}, the Kreith model based on the principle of heat/mass transfer analogy (HMTA) in the diffusion boundary layer was used for calculation [13,29]:

$$\dot{m} = \frac{h_{\text{conv}}}{k} \frac{D_{\text{steam}} M_{\text{steam}} P}{RT} \frac{(P_{\text{steam,i}} - P_{\text{steam,g}})}{P_{\text{non,avg}}} \left(\frac{Sc}{Pr}\right)^{1/3} \tag{11}$$

where: $\dot{m}$—steam condensation rate per unit area;
k—thermal conductivity of mixture in diffusion boundary layer;
D_{steam}—steam diffusion coefficient;
M_{steam}—molar mass of steam;
P—total mixture pressure;
$P_{\text{steam,i}}$—partial pressure of steam at phase interface;
$P_{\text{steam,g}}$—partial pressure of steam in main flow;
$P_{\text{non,avg}}$—mean partial pressure of noncondensable gas in gas phase;
Sc—Schmidt number.

In calculation, it is assumed that the steam at the interface is saturated, and that film condensation occurs on the wall surface. After condensation, the liquid film is uniformly attached to the wall surface. At the same time, the thickness of the film can be expressed by combining the condensing mass quality and density of the condensed water and the heat transfer area of the wall surface.

After calculating the condensation rate, the heat transfer coefficient of condensation is further calculated as follows:

$$q_{\text{cond}} = \dot{m} h^{*}_{\text{fg}} \tag{12}$$

$$h_{\text{cond}} = \frac{q_{\text{cond}}}{T_{\text{g}} - T_{\text{i}}} = \frac{\dot{m} h^{*}_{\text{fg}}}{T_{\text{g}} - T_{\text{i}}} \tag{13}$$

For the total heat transfer coefficient h_{tot}, since the calculations of h_{film}, h_{conv} and h_{cond} all require the interface temperature T_{i} in order to obtain the total heat transfer coefficient h_{tot}, the liquid film heat transfer coefficient, convective heat transfer coefficient and steam condensation heat transfer coefficient need to be calculated successively by assuming the initial value of the liquid film surface temperature, and then a new T_{i} is obtained through the equations above, and an iteration is performed until the difference between the new and old interface temperatures T_{i} reaches an acceptable error; then, the final interface temperature T_{i} is obtained. After that, the total heat transfer coefficient h_{tot} can be obtained.

3. Verification and Validation of Condensation Heat Transfer Model

After the construction of the condensation model, it is necessary to carry out further verification and validation of the model. For verification, this paper uses the code-to-code method to model and calculate the same example by using the PISAA and traditional thermal hydraulic codes, respectively, and then compares the calculated results from the different codes. For validation, this paper creates a calculation example by referring to the Wisconsin condensation experiment, and then compares the calculated results with the experimental data.

3.1. Model Verification Using Traditional Containment Analysis Codes

By utilizing the developed heat transfer calculation model of the internal heat structure of the containment, which is implemented in the autonomous integral analysis code, PISAA, the calculation is performed using a control volume coupled with a heat structure. If the coupling between the control volume and the heat structure is in a non-equilibrium state, the heat structure serves as a heat sink and transfers heat and mass with the internal fluid until the system reaches a steady state. During the process, parameters such as the composition and state of the fluid inside the control volume, and the position, direction and boundary conditions of the heat structure are all important factors that affect the system's ability to reach its final equilibrium state and should be taken into account, respectively, in calculation. Moreover, the developed computing code is validated by combining the parameters with the mainstream containment computing code and conducting the same case calculations. The conditions for the calculation are shown in Table 1 below.

Table 1. Initial testing conditions of control volume coupled with heat structure.

Parameters	Value
Volume (m^3)	1000
Elevation (m)	0.0
Control volume height (m)	15.0
Hydraulic diameter (m)	9.2132
Pressure (kPa)	101.0
Gas phase temperature (K)	313.15
Liquid phase temperature (K)	-
Relative humidity (%)	0.1
Liquid phase fraction (%)	0.0
Surface area (m^2)	552.0
Elevation (m)	0.0
Heat structure height (m)	5.0
Thickness (m)	0.3048
Direction	Vertical
Boundary condition A	Convective
Boundary condition B	Convective
Initial temperature (K)	353.15
Calculation time (s)	15,000.0

For the above case, calculations were carried out using traditional containment analysis codes, code A, code B and PISAA, respectively. The pressure and gas temperature in the control volume, as well as the wall temperature and heat transfer coefficient of the heat structure, were analyzed sequentially, and the comparison diagrams are shown in Figure 2 below.

Figure 2 shows the variations in the pressure and gas temperature of the control volume, and the wall temperature of the heat structure when it is vertically placed inside the control volume. It can be observed from the figure that, after the process starts, heat transfer occurs between the fluid and the heat structure in the control volume due to the heat disequilibrium. The heat release from the high-temperature heat structure causes increases in the temperature and pressure of the gas inside the control volume, which gradually stabilize over time. Since the heat storage capacity of the gas in the control volume is much smaller than that of the heat structure, the temperature rise in the gas is much larger than the wall temperature of the heat structure. As can be seen from the figure, the temperature of the gas increases by about 40 K, while the wall temperature of the heat structure finally changes by less than 0.1 K.

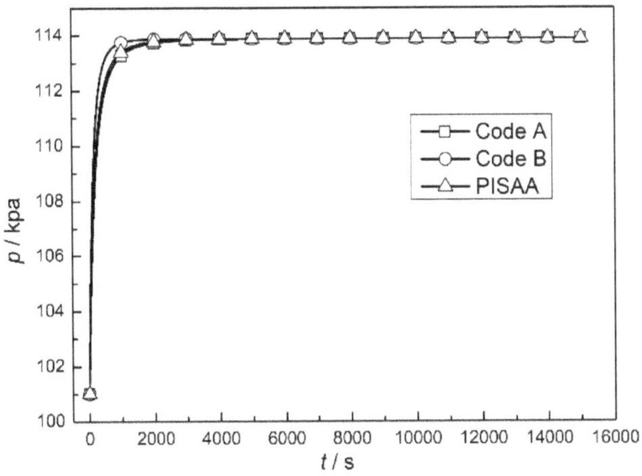

(**a**)

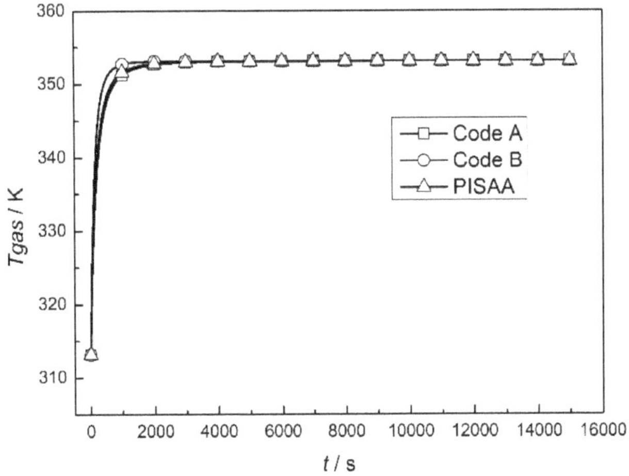

(**b**)

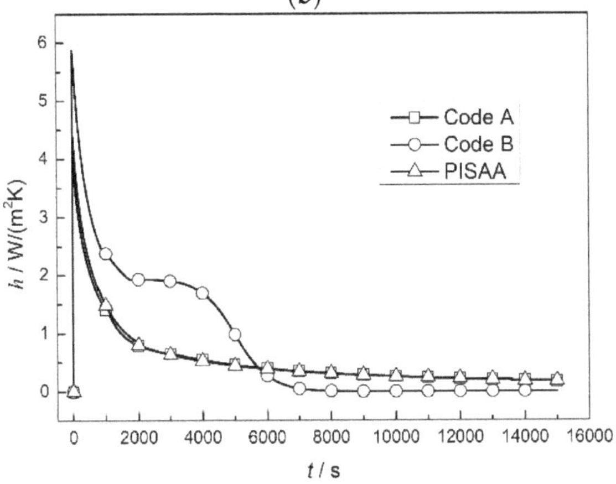

(**c**)

Figure 2. *Cont.*

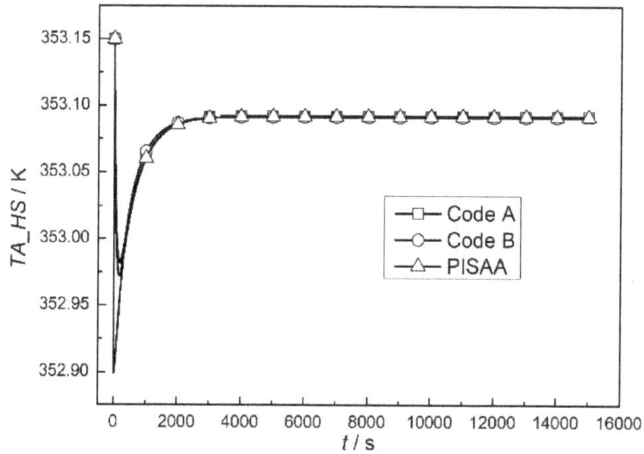

(d)

Figure 2. Comparison of calculation results of control volume coupled heat structure. (**a**) Pressure. (**b**) Gas temperature. (**c**) Wall heat transfer coefficient. (**d**) Wall temperature.

The convective heat transfer coefficient increases significantly after the beginning of the process, because the temperature difference between the gas temperature and the wall temperature is the largest at the beginning. For the natural convection heat transfer process in a limited space, the corresponding Grashof number is the largest at this time. Meanwhile, since the qualitative temperature of the fluid is half of the sum of the gas temperature and the wall temperature of the heat structure, it is also the maximum value. As the temperature of the gas approaches the temperature of the heat structure, the heat transfer coefficient tends to be stable.

As the system gradually approaches stability, there is a certain deviation in the calculation results of the convective heat transfer coefficient for three codes. But the change trend is basically consistent and there is only a small variation in the steady-state values. The results of the model used in this paper are more similar to the results obtained from code A. After the internal parameters of the system reach a steady state, their values generally remain unchanged.

3.2. Model Validation via Wisconsin Condensation Experiment

In addition, to validate the surface condensation heat transfer model for the heat structure, the results of this paper are compared with the Wisconsin atmospheric partial condensation experimental data [15]. The Wisconsin experiment mainly studied the condensation heat transfer phenomenon in the presence of noncondensable gases. The experiment considered condensation on both vertical and horizontal flat plates in a large space. Six aluminum condensation plates, with a total length of 0.9144 m and a thickness of 0.3048 m, were installed on the top and side walls located in the upper-right corner of the container. The container was filled with humid air maintained at a constant temperature and pressure. The humid air led to the condensation of heat on the wall of the aluminum condensation plates. During the experiment, the external surface of the condensation plate was cooled using continuously subcooled water to keep the plate wall temperature constant inside the large container, and high-temperature steam was injected from the bottom of the container to maintain the initial constant temperature and pressure state under different operating conditions. By setting different initial thermal parameters, the condensation heat transfer coefficients were measured under different experimental conditions using the HFM (Heat Flux Meter) and CEB (Coolant Energy Balance) methods. Table 2 briefly lists the necessary initial parameters for each experimental condition.

Table 2. Initial condition of condensation heat transfer.

No.	Pressure (kPa)	Molar Fraction of Noncondensable Gas (%)	Wall Temperature (°C)	Gas Temperature (°C)
1	100.0	0.795	28.60	60.65
2	100.0	0.699	29.40	69.23
3	100.0	0.702	29.40	69.04
4	100.0	0.713	29.20	68.15
5	100.0	0.533	34.00	79.68
6	100.0	0.420	28.43	85.12
7	100.0	0.417	29.76	85.25
8	100.0	0.433	29.47	84.55
9	100.0	0.312	30.60	89.53
10	100.0	0.307	30.30	89.72
11	100.0	0.308	29.96	89.68
12	100.0	0.498	27.76	81.46
13	100.0	0.511	31.74	80.80
14	100.0	0.516	29.58	80.58

In the Wisconsin experiment, two different methods were used to measure the convective heat transfer coefficient, but the uncertainty of the two measurement results was not given. Therefore, in order to improve the reliability of the experimental comparison results, the average value of the results obtained from the two measurement methods was used.

During the modeling process, one side surface of the heat structure was subjected to convection heat transfer, while the other side was kept at a constant temperature, and the air with steam was condensed on the heat structure surface. Figure 3 shows the comparison of the heat transfer coefficients between the experimental data and calculation results of the condensation for the vertical and horizontal condensing plates, respectively.

According to the initial parameters, it can be known that the mixed steam containing noncondensable gas will condense on the condensing plate. As can be seen from Figure 3, the condensation heat transfer coefficient calculated via the PISAA is basically consistent with the experimental data for all cases. However, for the conditions with relatively more noncondensable gas such as cases 1, 2, 3 and 4, the heat transfer coefficient is significantly lower than the other conditions. Taking conditions 1 and 6 of the vertical condensing plate as examples, when the molar fraction of noncondensable gas decreases from 0.795 to 0.42, the heat transfer coefficient increases from 75.8 W/m^{2}K to 238.9 W/m^{2}K. The heat transfer coefficient increases obviously with the decrease in noncondensable gas, which further indicates that noncondensable gas has a great influence on heat transfer.

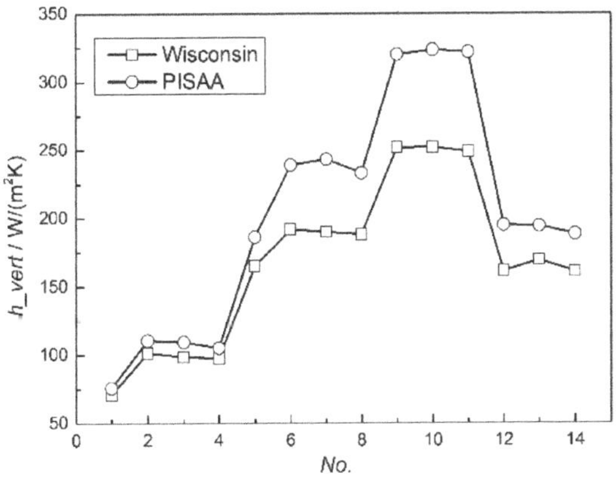

(a)

Figure 3. *Cont.*

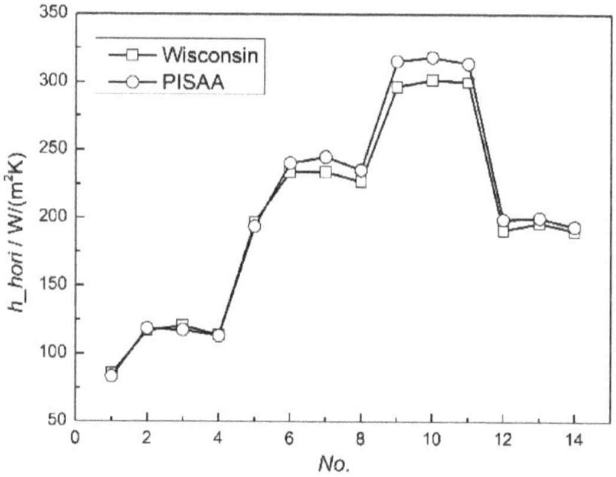

(**b**)

Figure 3. Comparison of condensation heat transfer coefficient between experimental data and calculation result. (**a**) Vertical condensing wall. (**b**) Horizontal condensing wall.

The calculation results for the vertical condensing plate in cases 9, 10 and 11 show a slightly larger deviation from the experimental data, with the maximum difference of 29.1477%, but are still within the acceptable deviation range of engineering. The other cases show smaller deviations. The average error of the heat transfer coefficient of horizontal condensation is about 2.5481%, which is in good agreement with the experimental data. The comparison results indicate that the wall condensation model used in the PISAA can simulate the physical process of wall condensation scientifically and accurately.

4. Parameter Sensitivity Analysis

After the verification and validation of the condensation model, it is necessary to carry out an appropriate analysis for the sensitivity parameters that affect the calculation results. There are many factors that affect the condensation heat transfer coefficient of a heat structure. The setting of the system parameters and the selection of the calculation models can both cause fluctuations in the results. In order to further investigate the accuracy and reliability of the model, this paper conducted a sensitivity analysis of the code by varying the geometric parameters of the system and calculating models, respectively.

4.1. Mesh Node Independence in Heat Structure

The node division of a heat structure affects the surface temperature to a certain extent. The denser the node division, the closer the node temperature to the real situation, which will then affect the physical property parameters of the fluid. For example, in the calculation of the convective heat transfer coefficient in Equations (8) and (9), the *Re*, *Gr* and *Pr* numbers all depend on the physical property parameters, such as the specific heat and viscosity, which may have a certain influence on the calculation results. Therefore, the number of nodes divided in the heat structure is analyzed as a sensitive parameter in this paper.

Tables 3 and 4 provide a detailed analysis of the influence of node numbers on the heat transfer coefficient of the condensing plate in different directions when they are divided. In this paper, the heat structure was divided into two, five and ten nodes along the heat transfer direction, respectively. After the calculation was completed, the comparison of the results revealed that the heat transfer coefficient of the heat structure's wall was hardly influenced by the number of nodes, and the deviation of the calculation results for each case was within 5‰, which can be considered negligible. In fact, because the wall thickness of the condensing plate is small, the overall thermal conductivity and thermal resistance is small, which can make the nodes in the heat structure reach the thermal equilibrium state in a very short time; therefore, the differences caused by thermal diffusion can be completely ignored. Consequently, the number of nodes in the heat structure will not significantly affect the magnitude and range of the heat transfer coefficient.

Table 3. Influence of the number of nodes on heat transfer coefficient for a vertical condensing plate.

No.	Wisconsin	2 Nodes	5 Nodes	10 Nodes
1	70.8	75.87777	75.87933	75.8811
2	101.38	110.402	110.4073	110.413
3	98.47	109.2021	109.2072	109.212
4	97.52	105.0613	105.0658	105.071
5	165.07	186.3355	186.3552	186.374
6	191.92	238.8905	238.9315	238.97
7	189.88	243.1433	243.1854	243.225
8	188.22	232.9544	232.9926	233.028
9	251.88	319.8702	319.9486	320.021
10	251.89	323.3353	323.4157	323.491
11	249.07	321.6682	321.7479	321.822
12	161.42	194.8331	194.8586	194.883
13	169.34	194.1343	194.1583	194.181
14	161.015	188.3783	188.4013	188.423

Table 4. Influence of the number of nodes on heat transfer coefficient for a horizontal condensing plate.

No.	Wisconsin	2 Nodes	5 Nodes	10 Nodes
1	85.28	83.3033	83.30575	83.3083
2	116.935	118.4498	118.4575	118.4649
3	120.45	117.2453	117.2528	117.26
4	113.645	113.0738	113.0805	113.0871
5	196.605	193.7196	193.7462	193.7713
6	233.79	240.1341	240.1856	240.2329
7	233.895	244.776	244.8291	244.8777
8	226.695	235.2196	235.268	235.3124
9	296.38	315.387	315.4822	315.5691
10	301.405	318.3065	318.4039	318.493
11	299.725	313.5457	316.6421	316.7304
12	190.97	199.0317	199.0647	199.0952
13	196.47	200.126	200.1576	200.1863
14	190.47	193.7491	193.7791	193.8068

4.2. Calculation Model for Mass Diffusivity

The physical parameters of water, steam, noncondensable gases and various materials are repeatedly used in the calculations, so the calculated values of these physical parameters affect the accuracy of the results to some extent. It can be seen from Equation (11) that the value of the mass diffusion coefficient D_{AB} has an impact on the mass condensation rate. Different calculation models of the mass diffusion coefficient will cause a disturbance to the condensing amount of steam per unit area.

According to the principle of the heat/mass transfer analogy, combined with Equations (12) and (13), it can be seen that the condensing heat transfer rate is also further affected by D_{AB}. Therefore, a sensitivity analysis of the mass diffusion coefficient is performed here.

When representing the mass diffusivity of steam to other mixed noncondensable gases, the Wilke and Lee model and the Fuller model were selected in this paper to calculate the mass diffusion rate of steam to a single noncondensable gas [30], and after obtaining the binary mass diffusivity, the final calculation result was obtained using Wilke's model for the mass diffusivity of steam to mixed gases.

The Wilke and Lee model is expressed as

$$D_{\mathrm{AB}} = \frac{\left[3.03 - \left(0.98/M_{\mathrm{AB}}^{1/2}\right)\right]\left(10^{-3}\right)T^{3/2}}{PM_{\mathrm{AB}}^{1/2}\sigma_{\mathrm{AB}}^{2}\Omega_{\mathrm{D}}} \tag{14}$$

The Fuller model is expressed as

$$D_{\mathrm{AB}} = \frac{0.00143T^{1.75}}{PM_{\mathrm{AB}}^{1/2}\left[\left(\sum_{\mathrm{v}}\right)_{A}^{1/3} + \left(\sum_{\mathrm{v}}\right)_{B}^{1/3}\right]^{2}} \tag{15}$$

where: D_{AB}—mass diffusion coefficient;

P—total gas pressure;

T—gas temperature;

M—molar mass.

Parameter values such as σ_{AB}, Ω_{D} and $\sum_{\mathrm{v}}$ can be obtained by looking up the parameter table [30].

By selecting two different binary mass diffusivity calculation models, the final heat transfer coefficient obtained from the code was compared with the experimental data from the Wisconsin condensation heat transfer experiment, as shown in Figure 4. The results show that although there is a significant difference between the condensation heat transfer coefficient calculated using the code and the experimental data, the results calculated using different mass diffusivity models are indeed different, and the deviation is about 5–10%. The results from the Wilke and Lee model were significantly closer to the experimental data than those from the Fuller model, indicating that physical parameters have an effect on the accuracy of the calculated results.

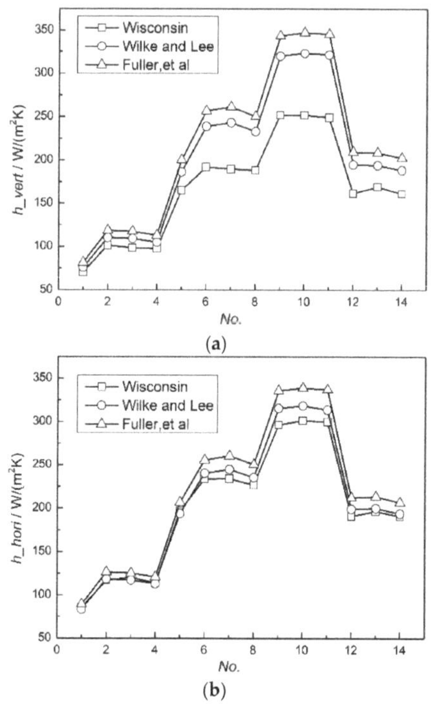

Figure 4. Influence of mass diffusivity model on condensation heat transfer coefficient. (**a**) Vertical condensing wall. (**b**) Horizontal condensing wall.

4.3. Condensation Model

The condensation phenomenon of steam containing noncondensable gas plays an important role in the heat removal process of the containment in a nuclear power plant under an accident condition. As can be seen from Equation (2), the total heat transfer coefficient is affected by both the convective heat transfer coefficient and the condensation heat transfer coefficient, and there are certain differences in the calculation results obtained using the different condensation calculation models. So, it is important to select an accurate condensation model that is suitable for the development of a severe accident analysis code. In this paper, the most widely used experimental correlation models of Uchida [7], Tagami [8] and Dehbi [9] are selected successively to compare with the experimental results and the calculated values of the PISAA code model.

$$h_{\text{Uchida}} = 380.0\left(\frac{W_{\text{nc}}}{1-W_{\text{nc}}}\right)^{-0.7} \tag{16}$$

$$h_{\text{Tagami}} = 11.4 + 284\left(\frac{1-x_{\text{nc}}}{x_{\text{nc}}}\right) \tag{17}$$

$$h_{\text{Dehbi}} = \frac{L^{0.05}[3.7+28.7p-(2483+458.3p)log_{10}W_{\text{nc}}]}{\left(T_g-T_w\right)^{0.25}} \tag{18}$$

where: h—condensing heat transfer coefficient; W_{nc}—noncondensable gas mass fraction; x_{nc}—noncondensable gas volume fraction; L—characteristic length of condensing wall; p—pressure; T_g—gas temperature; T_w—condensing wall temperature.

The comparison results are shown in the Figure 5.

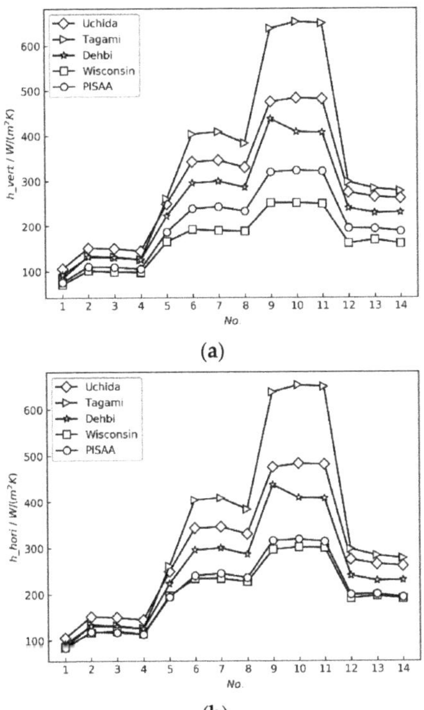

Figure 5. Effect of condensation model on heat transfer coefficient of condensation. (**a**) Vertical condensation wall. (**b**) Horizontal condensation wall.

The comparison shows that the calculation results of the condensation model based on Fick's law and the heat/mass transfer analogy principle chosen in this paper are closer to the experimental measurement data from the Wisconsin experiment. The deviation between the calculation results from the Tagami model and the experimental data is the largest, and the accuracy of the Dehbi model is slightly higher than that of Tagami and Uchida, but the deviation is much larger than that of the calculation model used in this paper.

To analyze the reason, the Uchida and Tagami models only consider the influence of the noncondensable gas fraction on the condensation heat transfer coefficient, and the Dehbi model is further improved, considering the influences of wall length, total pressure and wall subcooling degree on the condensation heat transfer coefficient, but the applicability range of the model is limited, and the factors affecting the mass and heat transfer of condensation, such as the temperature and composition of the gas in the main flow and the interface, are not considered in detail. The calculation model used in the PISAA code considers all the factors mentioned above; therefore, the calculation results are closer to the actual experimental measurements, and at the same time, simulate better.

The comparison results show that the condensation heat transfer model used in this paper, which is based on simplifying and integrating the mechanism and the experimental correlations, can effectively solve the limitations of the traditional empirical correlations model, and realize the accurate simulation of the condensation heat transfer of steam with noncondensable gas, and the calculation results are more reliable than the results calculated using the pure empirical correlations.

5. Conclusions

Based on the law of energy conservation and the thermal resistance relationship of each heat transfer process, combined with the mechanism and experimental correlations, this paper integrates a comprehensive calculation model of the condensation heat transfer of steam with noncondensable gas, including convection heat transfer, liquid film heat transfer and steam condensation heat transfer, and also considers the influence of various factors such as the condensation amount and flow patten classification in the liquid film on the process. The total heat transfer coefficient is obtained via the iterative method, the model is more applicable under different conditions and the calculated results are closer to the experimental data.

The condensation model is then used in an analysis code, PISAA, and the model is verified by comparing the calculation results with those from traditional containment analysis codes towards the same example, the change trend is basically consistent and the final steady-state values are only in a small variation range. And in the calculation and validation of the condensation heat transfer experiment in Wisconsin using the PISAA software, the average deviation of the heat transfer coefficient of the horizontal condensation heat transfer is 2.55%, and the maximum calculation deviation of the vertical condensing plate is about 29%, which is within the acceptable range of engineering. The reason for the large deviation may be that the simplified method of condensate film thickness in the vertical wall is not consistent with the actual situation, which can be further optimized and improved in the future.

Through the analysis of the sensitive parameters affecting the heat transfer coefficient, the division of the nodes on the condensing plate is less, that is, the temperature distribution has a little affect on the heat transfer coefficient. For the mass diffusivity calculation model, the maximum deviations between the calculated results obtained using the Wilke and Lee models and the Fuller model and the experimental values are 29% and 36%, respectively, and the physical property parameters affect the accuracy of the calculated results to some extent.

The deviations from different condensation models' calculated results and the experimental data are much larger than that of the condensation model used in this paper, indicating that the model is scientific and reasonabl, and can be used for thermal hydraulic analysis in a nuclear power plant.

The prototypes of the calculation equations in this paper come from the existing experimental data and references, and are simplified according to the theory and practical situation, integrated and coupled together in the calculation code, PISAA, which solves the problem of calculating the condensation of high-temperature and high-pressure gas in the containment in a nuclear power plant under a severe accident scenario, and has great significance for the accident analysis and simulation of a nuclear power plant and the improvement in the nuclear power safety level.

Author Contributions: Conceptualization, Y.Y. and X.Y.; methodology, X.Y. and R.M.; software, H.L. and S.S.; formal analysis, C.W.; investigation, H.L.; writing—original draft preparation, H.L. and S.S.; writing—review and editing, H.L. and C.W.; supervision, R.M.; project administration, Y.Y. All authors have read and agreed to the published version of the manuscript.

Funding: This research was funded by the National Key R&D Program of China (2019YFB1900704) grant, number 2.2 million.

Institutional Review Board Statement: Not applicable.

Informed Consent Statement: Not applicable.

Data Availability Statement: The data presented in this study are available upon request from the corresponding authors.

Conflicts of Interest: The authors declare no conflict of interest.

Nomenclature

T	temperature, K
q	heat flux, W/m^2
h	heat transfer coefficient, $W/(m^2\ K)$
D	diffusion coefficient, m^2/s
M	molar mass, kg/mol
R	gas constant, J/(mol K)
k	thermal conductivity, W/(m K)
Sc	Schmidt number
Pr	Prandtl number
X	gas volume fraction, %
L	characteristic length, m
W	mass fraction, %
g	gas
l	liquid
steam	steam
w	wall
nc	noncondensable gas

References

1. Dela, R.J.; Escriva, A.; Herranz, L.E.; Cicero, T.; Munoz-Cobo, J.L. Review on condensation on the containment structures. *Prog. Nucl. Energy* **2009**, *51*, 32–66.
2. Yadav, M.K.; Khandekar, S.; Sharma, P.K. An integrated approach to steam condensation studies inside reactor containments: A review. *Nucl. Eng. Des.* **2016**, *300*, 181–209.
3. Huang, J.; Zhang, J.X.; Wang, L. Review of vapor condensation heat and mass transfer in the presence of non-condensable gas. *Appl. Therm. Eng.* **2015**, *89*, 469–484. [CrossRef]
4. Su, J.Q.; Fan, L.; Gao, L. Review of steam condensation heat transfer under containment cooling system condition. *At. Energy Sci. Technol.* **2016**, *50*, 1956–1966.
5. Colburn, A.P.; Hougen, O.A. Design of cooler condensers for mixtures of vapors with noncondensing gases. *Ind. Eng. Chem.* **1934**, *26*, 1178–1182. [CrossRef]
6. Hasson, D.; Luss, D.; Navon, U. An experimental study of steam condensation on a laminar water sheet. *Int. J. Heat Mass Transf.* **1964**, *7*, 983–1001. [CrossRef]
7. Uchida, H.; Oyama, A.; Togo, Y. Evaluation of Post—Incident Cooling Systems of Light Water Power Reactors. In Proceedings of the Third International Conference on the Peaceful Uses of Atomic Energy, Geneva, Switzeland, 31 August–9 September 1964.

8. Tagami, T. *Interim Report on Safety Assessments and Facilities Establishment Project for June 1965*; Japanese Atomic Energy Research Agency: Tokyo, Japan, 1965.
9. Dehbi, A.A. *The Effects of Noncondensable Gases on Steam Condensation under Turbulent Natural Convection Conditions*; Massachusetts Institute of Technology: Cambridge, MA, USA, 1991.
10. Fan, G.M.; Tong, P.; Sun, Z.N.; Chen, Y. Development of a new empirical correlation for steam condensation rates in the presence of air outside vertical smooth tube. *Ann. Nucl. Energy* **2018**, *113*, 139–146. [CrossRef]
11. Bhowmik, P.K.; Schlegel, J.P.; Revankar, S. State-of-the-art and review of condensation heat transfer for small modular reactor passive safety: Experimental studies. *Int. J. Heat Mass Transf.* **2022**, *192*, 122936. [CrossRef]
12. Murase, M.; Kataoka, Y.; Fujii, T. Evaporation and condensation heat transfer with a non-condensable gas present. *Nucl. Eng. Des.* **1993**, *141*, 135–143. [CrossRef]
13. Ambrosini, W.; Forgione, N.; Manfredini, A.; Oriolo, F. On various forms of the heat and mass transfer analogy: Discussion and application to condensation experiments. *Nucl. Eng. Des.* **2006**, *236*, 1013–1027. [CrossRef]
14. Oh, S.; Revankar, S.T. Experimental and theoretical investigation of film condensation with noncondensable gas. *Int. J. Heat Mass Transf.* **2006**, *49*, 2523–2534. [CrossRef]
15. Anderson, M.H. Steam Condensation on Cold Walls of Advanced PWR Containments. Ph.D. Thesis, The University of Wisconsin, Madison, WI, USA, 1998.
16. Zhang, L.L.; Zhang, G.M.; Tian, M.C.; Zhang, J.; Zhang, Y. Modeling of laminar filmwise condensation of methane with nitrogen on an isothermal vertical plate. *Int. Commun. Heat Mass Transf.* **2019**, *105*, 10–18. [CrossRef]
17. Wang, T.L.; Wang, D.; Tong, L.L.; Cao, X. Improvement of Condensation Model With the Presence of Non-Condensable Gas for Thermal-Hydraulic Analysis in Containment. *Front. Energy Res.* **2021**, *9*, 671539. [CrossRef]
18. Hammoudi, D.; Benabdesselam, A.; Azzi, A.; Kassim, M.A. Numerical modeling of steam condensation in vertical channel in presence of noncondensable gas. *Int. J. Therm. Sci.* **2018**, *126*, 263–271. [CrossRef]
19. Li, S.W.; Wei, M.; Wang, X.J. A numerical study on film condensation of steam with non-condensable gas on a vertical plate. *Front. Energy Res.* **2023**, *11*, 1064067. [CrossRef]
20. Vyskocil, L.; Schmid, J.; Macek, J. CFD simulation of air–steam flow with condensation. *Nucl. Eng. Des.* **2014**, *279*, 147–157. [CrossRef]
21. Punetha, M.; Khandekar, S. A CFD based modelling approach for predicting steam condensation in the presence of non-condensable gases. *Nucl. Eng. Des.* **2017**, *324*, 280–296. [CrossRef]
22. Liu, F.; Sun, Z.N.; Cao, B.Y.; Bian, H.; Ding, M. Numerical investigations on the component separation phenomenon and transport behavior of steam-air-hydrogen mixture induced by condensation. *Int. J. Therm. Sci.* **2022**, *172*, 107369. [CrossRef]
23. Wen, J.; Liang, M.Z.; Zhang, H.W.; Cheng, X.H. Numerical simulation of steam condensation in vertical tube with the presence of non-condensable gas. *J. Shanghai Jiaotong Univ.* **2009**, *43*, 299–303.
24. Sun, S.H.; Li, S.G.; Shaheen, S.; Arain, M.B.; Usman; Khan, K.A. A numerical investigation of bio-convective electrically conducting water-based nanofluid flow on the porous plate with variable wall temperature. *Numer. Heat Transf. Part A Appl.* **2023**, 1–15. [CrossRef]
25. Peng, X.; Cao, X.X.; Cao, J.H.; Ding, M.; Bian, H. Numerical study on the effect of CO_2 on the heat transfer characteristics of steam condensation outside a vertical tube. *Nucl. Eng. Des.* **2022**, *391*, 111745. [CrossRef]
26. Ma, X.Z. Research on Steam Condensation with non -Condensable Gases under the High Pressure. Ph.D. Thesis, Tsinghua University, Beijing, China, 2017.
27. Ganguli, A.; Patel, A.G.; Maheshwari, N.K.; Pandit, A.B. Theoretical modeling of condensation of steam outside different vertical geometries (tube, flat plates) in the presence of non-condensable gases like air and helium. *Nucl. Eng. Des.* **2008**, *238*, 2328–2340. [CrossRef]
28. Liu, Z.; Jin, H.X.; Li, X.Y.; Wu, X. Research on condensation heat transfer in the presence of noncondensable gases based on Fick's law and HMTA. *Nucl. Saf.* **2017**, *16*, 85–90.
29. Bergman, T.L.; Lavine, A.S.; Frank, P.; DeWitt, D.P. *Fundamentals of Heat and Mass Transfer*, 7th ed.; John Wiley & Sons: Hoboken, NJ, USA, 2011.
30. Bruce, E.P.; John, M.P.; John, P.O. *The Proper Ties of Gases and Liquids*, 5th ed.; The McGraw-Hill Companies, Inc.: New York, NY, USA, 2001.

MDPI
St. Alban-Anlage 66
4052 Basel
Switzerland
www.mdpi.com

Applied Sciences Editorial Office
E-mail: applsci@mdpi.com
www.mdpi.com/journal/applsci

www.ingramcontent.com/pod-product-compliance
Lightning Source LLC
Chambersburg PA
CBHW062052220926
51850CB00010B/2749
* 9 7 8 3 7 2 5 8 1 1 3 9 7 *